METHODS IN INHALATION TOXICOLOGY

Edited by

Robert F. Phalen, Ph.D.

CRC Press
Boca Raton New York London Tokyo

CRC Press
METHODS IN THE LIFE SCIENCES

Gerald D. Fasman - Advisory Editor
Brandeis University

Series Overview

Methods in Biochemistry
John Hershey
Department of Biological Chemistry
University of California

Cellular and Molecular Neuropharmacology
Joan M. Lakoski
Department of Pharmacology
Penn State University

Research Methods for Inbred Laboratory Mice
John P. Sundberg
The Jackson Laboratory
Bar Harbor, Maine

Methods in Neuroscience
Sidney A. Simon
Department of Neurobiology
Duke University

Joseph M. Corless
Department of Cell Biology,
Neurobiology and Ophthalmology
Duke University

Methods in Pharmacology
John H. McNeill
Professor and Dean
Faculty of Pharmaceutical Science
The University of British Columbia

Methods in Signal Transduction
Joseph Eichberg, Jr.
Department of Biochemical and Biophysical Sciences
University of Houston

Methods in Toxicology
Edward J. Massaro
Senior Research Scientist
National Health and Environmental Effects Research Laboratory
Research Triangle Park, North Carolina

CRC Press
METHODS IN TOXICOLOGY

***Edward J. Massaro** - Advisory Editor*
Senior Research Scientist
National Health and Environmental Effects Research Laboratory
Reasearch Triangle Park, North Carolina

The **CRC Press Methods in Toxicology Series** provides the reader with a step-by-step approach to each of the up-to-date methods and techniques which are presented in a clear and concise format. Topics covering all aspects of the methods that impact on toxicology are being reviewed for publication.

Published Titles

Methods in Renal Toxicology, Rudolfs K. Zalups and Lawrence H. Lash

Molecular and Cellular Methods in Developmental Toxicology, George P. Daston

Methods in Inhalation Toxicology, Robert F. Phalen

Forthcoming Titles

PCR Application to Molecular Toxicology, Jack Vanden Heuvel

Senior Editor:	Paul Petralia
Editorial Assistant:	Cindy Carelli
Project Editor:	Sarah Fortener
Assistant Managing Editor:	Gerry Jaffe
Marketing Manager:	Susie Carlisle
Direct Marketing Manager:	Becky McEldowney
Cover design:	Dawn Boyd
PrePress:	Kevin Luong
Manufacturing:	Sheri Schwartz

Library of Congress Cataloging-in-Publication Data

Methods in inhalation toxicology / edited by Robert F. Phalen.
p. cm. -- (CRC Press methods in toxicology)
Includes bibliographical references and index.
ISBN 0-8493-3343-1
1. Pulmonary toxicology--Animal models. 2. Air--Pollution--Toxicity testing. I. Phalen, Robert F., 1940- . II. Series.
[DNLM: 1. Respiratory System--drug effects. 2. Toxicology--methods. 3. Air Pollutants--toxicity. WF 100 M592 1996]
RC720.M48 1996
615.9′02--dc20
DNLM/DLC
for Library of Congress 96-25496
CIP

International Standard Book Number 0-8493-3343-1
Library of Congress Card Number 96-25496
Printed in the United States of America 1 2 3 4 5 6 7 8 9 0
Printed on acid-free paper

Abstract

Today, and in the foreseeable future, an appropriate understanding of the biomedical effects of potentially significant natural and anthropogenic air contaminants will require conducting inhalation exposure studies employing whole living subjects. Conducting such studies is a specialized research activity that draws upon skills from a variety of scientific disciplines. How knowledge from these disciplines is integrated into the design and conduct of inhalation studies is the subject of this book, which was written by an experienced, active team of research scientists. Each of the vital aspects germane to the conduct of such studies is discussed in detail, including the selection of animal subjects and the related quality-control and ethical considerations, air purification methodology, exposure atmosphere generation and characterization, inhalation exposure systems, and real-time and post-exposure biological assessments. The book, which is targeted to students in toxicology programs and scientists contemplating performing inhalation studies, includes comprehensive lists and addresses of suppliers of useful equipment and materials relevant to each of the major aspects involved with the performance of valid, reproducible inhalation toxicology studies.

Preface

We live and breathe immersed in a sea of air. In addition to the major gaseous components, oxygen and nitrogen, countless additional gases, vapors, and particles are always present in the atmosphere and are available to deposit in the respiratory system. Some of these contaminants are natural, and some are anthropogenic. Of the anthropogenic substances, some are beneficial (such as aerosolized medications), and some are harmless or harmful depending upon their concentrations. If we are to maximize human wellbeing, we must study and understand the biomedical effects of each of the potentially significant air contaminants. Today, and in the foreseeable future, this study and understanding will require conducting inhalation studies using whole living subjects. In some studies, the subjects will be human volunteers, but in most studies laboratory animals will be involved. Conducting such studies is a specialized research activity that draws upon skills taken from numerous scientific disciplines. These disciplines include chemistry, physics, engineering, biology, physiology, microbiology, medicine, pathology, and others as well. How knowledge from these areas is integrated into the design and conduct of inhalation studies is the subject of this volume.

The chapters in this book were written by an active team of scientists who have worked together in a modern inhalation toxicology program at the Air Pollution Health Effects Laboratory at the University of California, Irvine. Performing such studies has been a major activity of this team for the past 10 years. In addition, these authors have trained students in research methods, and they have established and taught several formal courses related to the topics in this book. In short, each author is writing about his current daily activities and within the scope of his formal training and experience. It is hoped that this volume will provide a solid foundation for the readers — a foundation that will support further specialized learning and will lead to success in conducting their own inhalation studies.

In addition to the authors, several people have contributed significantly to the material herein. The chapters were reviewed and improved by the technical staff of the Air Pollution Health Effects Laboratory. Special thanks are due to Richard C. Mannix, M.S.; Charles Bufalino, M.S.; Laura A. Hoffman, B.A.; and Michael J. Oldham, M.S. Additional editing, checking, formatting, and word-processing were performed by Marie C. Tonini. Finally, this book is dedicated with gratitude and respect to the research subjects; they have been part of the team as well.

Robert F. Phalen, Ph.D.
Irvine, CA

The Editor

Robert Franklynn Phalen, Ph.D., directs the Air Pollution Health Effects Laboratory at the University of California, Irvine (UCI). He also holds two academic appointments in the College of Medicine at UCI: Professor in the Department of Community and Environmental Medicine and Professor in the Center for Occupational and Environmental Health (in the Department of Medicine). He is a member of the faculty for the Graduate Program in Environmental Toxicology at UCI, and he teaches medical students, medical residents, graduate students, and undergraduate students in courses at several universities including UCI, UCLA, and USC.

Dr. Phalen graduated in 1964 from San Diego State University with a B.S. in physics (and a mathematics minor) and from the same institution in 1966 with an M.S. degree in physics with a specialty in nuclear physics. In 1971, he obtained a Ph.D. in biophysics, with specialization in inhalation toxicology, from the University of Rochester (Rochester, NY). His postdoctorial research in aerosol physics and inhaled particle deposition modeling was conducted at the Inhalation Toxicology Research Institute in Alburquerque, NM.

In 1972, Dr. Phalen joined the College of Medicine at UCI to establish the Air Pollution Health Effects Laboratory, which has since conducted over 100 studies relating to the toxicology of air pollutants. These studies have been supported by a variety of federal agencies (including the U.S. EPA; the National Heart, Lung and Blood Institute; the National Institute of Environmental Health Sciences; and the Centers for Disease Control and Prevention), California state agencies (California Air Resources Board, the Department of Health Services, and others), and numerous private agencies (including the Electric Power Research Institute, the Health Effects Institute, the Charles C. Stocking Trust Fund, Southern California Edison Co., the Nickel Producers Environmental Research Association, and others). His research has covered several topics, including respiratory tract defenses, comparative respiratory tract anatomy, basic aerosol science, inhaled aerosol deposition models, particle and gas toxicity, and the inhalability of large particles.

Dr. Phalen is on the editorial board of three scientific journals, and was a guest editor for two. He has published over 100 scientific papers and authored a book entitled *Inhalation Studies: Foundations and Techniques* (published by CRC Press in 1984). He has made numerous invited presentations at international and national conferences and chaired sessions at same. Dr. Phalen has received many awards, including those from the American Legion, the U.S. Atomic Energy Commission, and the International Metallurgical Association; he recently received the Public Education Award from the California Biomedical Research Association. He is a member of ten professional and scientific associations relating to aerosol science, toxicology and industrial hygiene. He has two sons, Robert and Steven, and his wife, Katherine, is an occupational health nurse.

Contributors

Deepak K. Bhalla, Ph.D.
Associate Professor
Occupational and Environmental Health Sciences
Wayne State University
Detroit, Michigan

Michael T. Kleinman, Ph.D.
Professor
Air Pollution Health Effects Laboratory
Department of Community and Environmental Medicine
Center for Occupational and Environmental Health
College of Medicine
University of California
Irvine, California

William J. Mautz, Ph.D.
Associate Professor
Air Pollution Health Effects Laboratory
Department of Community and Environmental Medicine
College of Medicine
University of California
Irvine, California, *and*
Department of Biology
University of Hawaii
Hilo, Hawaii

Robert F. Phalen, Ph.D.
Professor
Air Pollution Health Effects Laboratory
Department of Community and Environmental Medicine
Center for Occupational and Environmental Health
College of Medicine
University of California
Irvine, California

Ronald Rasmussen, Ph.D.
Professor
Air Pollution Health Effects Laboratory
Department of Community and Environmental Medicine
College of Medicine
University of California
Irvine, California

Contents

Chapter 1

Introduction

Robert F. Phalen

Contents

0-8493-3343-1/97/$0.00+$.50

I. Routes of Exposure to Pollutants

A. Inhalation

Breathing produces by far the greatest total exposure of the body to environmental contaminants. Each breath brings at least a few hundred cubic centimeters (and up to several thousand cubic centimeters) of air into intimate contact with some of the most delicate membranes (the alveolar walls) in the body. When the inhaled (tidal) volume is multiplied by the number of breaths taken per minute and by the number of minutes in a day, one obtains daily intakes of air of 10,000 *l* (for an adult who never exercises) up to over 20,000 *l* (for a person who is physically very active much of the time). Compare these volumes with the average adult's daily intake of about 1.5 to 2 *l* of food and of water (Table 1.1). Also, as shown in the table, a similar situation occurs for children. Because the levels of contaminants in air, food, and water can be similar — in terms of mass of contaminant per liter — air easily can expose a person to about 10,000 times more mass of an environmental pollutant than does food or water.

TABLE 1.1
Intakes for an Adult and Child of the Major Media that Produce Exposure to Environmental Contaminants

	Daily intake (*l*)		Daily intake (g)	
Medium	Child (6 yr)	Adult	Child (6 yr)	Adult
Air				
Sedentary	4300	10,000	5200	12,000
Active	8000	20,000	9600	24,000
Water	0.6	1.5	600	1500
Food	0.2	0.6	200	600

The potential health consequences of the large daily intake of air in relation to other routes of pollutant exposure can be appreciated when one examines the commonly encountered air contaminants. Pure air is not found in nature; nearly pollutant-free air exists in clean-rooms that are built of inert materials and are supplied with highly filtered air under positive pressure to prevent inward leaks. Typical indoor or outdoor air that is not particularly polluted contains about 10 million fine particles per cubic meter and very low concentrations of hundreds of gases (other than oxygen and nitrogen) and vapors (in addition to water vapor). The pollutant load in air can easily be increased 100- to 1000-fold if an activity such as smoking, cooking, or metal-working is taking place. The full range of types of particulate contaminants

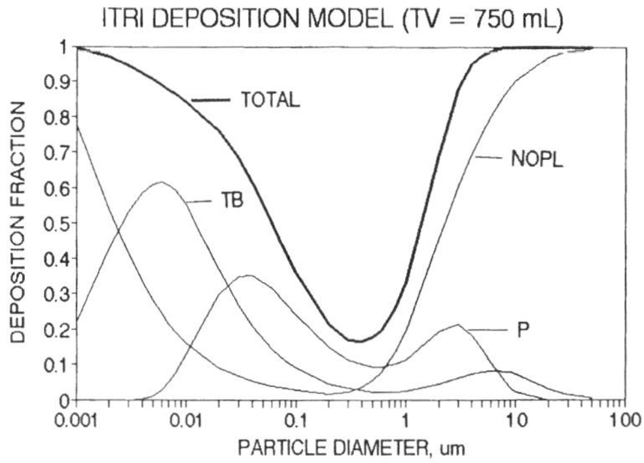

Figure 1.1
Particle deposition probabilities of unit density spherical particles inhaled by adults at a tidal volume of 750 m*l*. (Figure supplied by Dr. H.C. Yeh of the Inhalation Toxicology Research Institute, Albuquerque, NM.)

are too numerous to list completely, but the following are nearly always present in unfiltered air: viruses, bacteria, mold spores, pollen, dander (from humans and animals), insect and plant parts, soil dust, combustion products, metals and other inorganic substances, and organic compounds of various types. The expected probabilities of unit density (1 g/cm^3) particles of various sizes depositing in a normal, healthy adult's respiratory tract are shown in Figure 1.1.

It is surprising that the respiratory tract is able to survive this constant onslaught for very long. In fact, many of our illnesses are produced or contributed to by the aforementioned air contaminants. Not only do inhaled substances lead to diseases of the respiratory system, but also other organ systems of the body may be affected when the contaminants (or their metabolic products) are distributed via the flow of blood or lymph fluid. Examples of respiratory-tract disorders produced by air contaminants include: respiratory infections (colds, influenzas, and fugal diseases); inflammatory-type disorders such as bronchitis, rhinitis, asthma, and alveolitis; and chronic diseases such as emphysema, cancer, and fibrosis. A number of nonrespiratory system diseases that also may result from inhaled contaminants include bone diseases, liver diseases, nervous system diseases, and diseases of muscles and blood vessels. The fact that we are able to live long and relatively healthy lives is a credit to the many elaborate and effective defensive systems found throughout the body, including the respiratory tract.

Understanding inhalation phenomena, including the types of pollutants inhaled and their fates, is both a major and an important challenge of modern science. Through such an understanding we can expect to extend our productive lifespans and to help establish safe new technologies that improve the quality of life. In addition, scientific understanding of inhalation phenomena has many applications in medicine, because many medicines, vaccines, and diagnostic agents can be effectively and safely delivered by inhalation.

B. Other Routes of Exposure

Other routes of exposure to pollutants include ingestion (of food, beverages, medicines, and other substances), dermal (skin) absorption, exposure via wounds or injections, and even absorption through the eyes. Although we are not directly concerned with these nonrespiratory routes of exposure, air contaminants can play important roles in each of them. Air contaminants find their way into our food via inhalation by domesticated farm animals, and fallout from the air directly contaminates foodstuffs and water supplies that are consumed by these animals. Air pollution also affects water used for irrigating crops and for washing food. Contaminants can land on our food prior to consumption, also.

II. Respiratory Tract Anatomy

A. Human

Our current understanding of the structure of the human respiratory tract is far from complete. Considerable research has been focused on generating such data on the typical, or average, healthy adult; however, even these data are based on highly simplified structural templates, such as smooth cylindrical bronchial tubes, y-shaped airway bifurcations, uniformly sized spherical or polyhedral alveoli, and an absence of variation in the number of bronchial divisions required in a pathway to reach the alveoli. Even a brief examination of a replica cast of human lungs will reveal many deviations from these assumptions. Individual bronchial airways are seen to be variable in diameter, frequently noncircular, and often curved or bent between bifurcations. The branch points themselves can be sharp, blunt, or asymmetric or lead to three daughter airways rather than the usual two. Other variations in the usual structures are also present. In addition, little is known about age-related, race-related, gender-related, or disease-related effects on respiratory tract anatomy.

The principal anatomical structures of the respiratory tract include the: (1) nose, consisting of the nares, vestibule, and nasal cavity proper (with the conchae or turbinates); (2) nasopharynx; (3) oral cavity; (4) oropharynx; (5) laryngopharynx; (6) larynx; (7) trachea; (8) bronchi; (9) bronchioles; (10) respiratory bronchioles; (11) alveolar ducts; (12) alveolar sacs; and (13) alveoli. These structures are conveniently grouped into larger regions, or compartments, for the purpose of simplification and to facilitate mathematical modeling. A variety of compartmental schemes have been proposed, but several very similar models have been particularly useful to inhalation toxicologists. The models are those of Task Groups of the International Commission on Radiological Protection,[1-2] the Ad Hoc Working Group to Technical Committee 146–Air Quality–of the International Standards Organization,[3] the Air Sampling Procedures Committee of the American Conference of Governmental

TABLE 1.2
Compartmental Models of the Human Respiratory System as Developed by the ICRP Task Group on Lung Dynamics, the International Standards Organization, and the American Conference of Governmental Industrial Hygienists (ACGIH)

Region	Anatomic structures included	Task group region	ISO region	ACGIH region
1	Nose, mouth, nasopharynx, oropharynx, laryngopharynx, larynx	Nasopharynx (NP) or (NOPL)	Extrathoracic (E)	Head airways region (HAR)
2	Trachea, bronchi, bronchioles	Tracheobronchial (TB)	Tracheobronchial (B)	Tracheobronchial region (TBR)
3	Respiratory bronchioles, alveolar ducts, alveolar sacs, alveoli	Pulmonary (P)	Alveolar (A)	Gas exchange region (GER)

Industrial Hygienists,[4] and the Task Group of the National Council on Radiation Protection.[5] These compartmental systems, which are largely based on the pioneering morphometric work of Ewald Weibel,[6] are shown in Table 1.2 and Figure 1.2. Further information on the human respiratory tract models may be found in several references.[7-9]

B. Rat

The laboratory rat has structures similar to those of the human, except that the oral region need not be considered, as rats appear to nose-breathe only. Also, the transition from bronchioles to fully alveolarized ducts is so abrupt that the respiratory bronchioles are essentially absent. The airways of the rat differ from those of the human in several other ways, as well. The nasal region of the rat is much more complex due to the large posterior olfactory apparatus. The airway branching is much more asymmetric, and the number of generations of airways is much smaller. The airways of the rat are depicted in Figure 1.3. A quantitative model of the airway anatomy of the rat has been published by Yeh, Schum, and Duggan.[10]

Although this book is largely limited to methods that employ the laboratory rat, a great variety of other animals is used in inhalation studies. A special August 1983 issue of the *American Review of Respiratory Disease*, Vol. 128, No. 2 (Suppl.), entitled "Comparative Biology of the Lung", contains over 20 useful papers comparing several species.

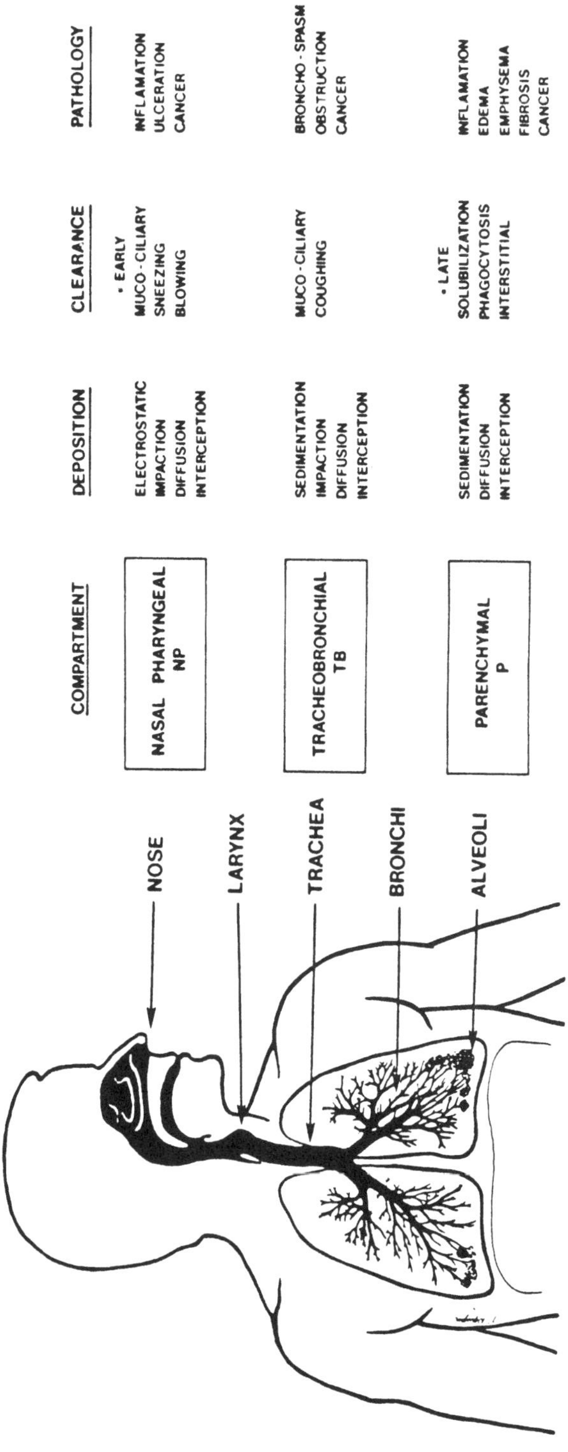

Figure 1.2
Compartmental model of the human's respiratory tract airways as defined by the Task Group on Lung Dynamics of the ICRP.[1]

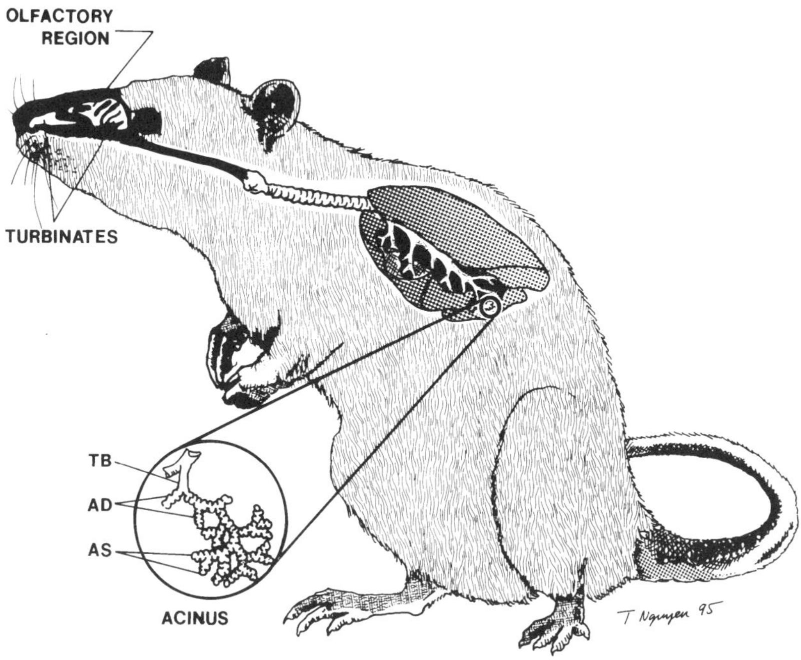

Figure 1.3

Airways of the laboratory rat. Location in the body is only approximate.

III. Major Types of Inhalation Studies

A. Toxicological

Fundamental to the practice of experimental toxicology is the principle that each route of exposure must be examined for its possible uniqueness. Many substances that are essentially harmless in food or water can produce serious illness and even death when inhaled in sufficient quantities. Among the myriad examples are many inorganic dusts; allergenic spores, pollens, and danders; several metals; certain microorganisms; and a variety of combustion products, including tobacco smoke. Also, the tissues of the respiratory system are capable of metabolizing (chemically transforming) many organic substances in ways that differ from their metabolism in the gastrointestinal tract. It is clear that inhalation studies frequently have no viable substitutes when one is investigating the consequences of breathing contaminated air.

Inhalation toxicology studies must be conducted with many types of air contaminants: natural, those produced by old or low-technology processes, and those produced by new or high-technology processes. Such studies require generating the test substance in physical and chemical forms that are as similar as is practical to the materials inhaled by humans. Also, when mixtures are present that may influence toxicity, the mixtures must be examined in inhalation studies. The journal *Inhalation Toxicology* publishes a variety of research studies that are of importance to inhalation toxicologists. Other journals that cover specialized related areas include the *American Industrial Hygiene Association Journal*, *Applied Occupational and Environmental Hygiene*, and *Toxicology and Environmental Health*.

B. Medical

Inhalation studies with medicinal substances must evaluate not only potential toxicity, but efficacy as well. In such studies, chemical and physical forms usually are intentionally varied in order to maximize the therapeutic or diagnostic effects. This additional requirement places demands on the flexibility of systems used to generate the particles and gases under study. Such demands are not to be taken lightly, as eventual approval of a medicinal agent can be limited to the exact physical forms (median particle size and size distribution, for example) that were used in the animal studies. *The Journal of Aerosol Medicine* publishes basic and applied research relating to the inhalation of medications.

C. Physiological

Inhalation studies are performed by scientists in their exploration of how living systems function. Particles in air can be used as tracers to follow clearance pathways or, alternatively, to learn how inhaled substances cross respiratory tract surfaces to gain access to the rest of the body. Physiologists also use inhalation studies to study the normal distribution of inhaled air throughout the lungs and within the nose and to examine the effectiveness of the mixing of air in successive breaths. Generally, such studies use particles and gases that, by design, have no toxic or physiologic effects of their own. The design and performance of such tracer studies have their own set of challenges. In addition to the previously mentioned journals, the *Journal of Applied Physiology* carries research papers on basic inhalation phenomena.

IV. Requirements of Inhalation Studies

From the preceding discussion it is clear that a particular study may have many unique and challenging requirements; however, some features of inhalation studies are essentially universal. An experimental atmosphere must be generated, controlled, characterized, and presented to experimental subjects. Unwanted contaminants must be effectively eliminated, and the air temperature and relative humidity must be within limits that do not stress the subjects or otherwise harm the experiment. The subjects themselves must be well characterized, appropriate for the study, and treated in an ethical manner (this is accomplished by minimizing stress, monitoring for pain or discomfort, medicating when needed, and even terminating the study, if necessary).

The implications of the requirements for successfully conducting a given inhalation study are substantial. First, a team of scientists may be required. This team typically includes a physical scientist (with aerosol or atmospheric science training), a physiologist (with pulmonary expertise), and a person who has primary expertise relating to the problem under study (toxicology, asthma, immunology, etc.). In addition, inhalation studies require specialized, sophisticated, well maintained, and, often, custom equipment. Air purification alone, which generally is essential to a modern inhalation study, can cost tens of thousands of dollars, take months of engineering design, and require a long-term commitment of funds and personnel for retrofits, maintenance, and repair. In addition, inhalation studies generally require much more space than do most biomedical investigations. Air moving equipment, animal isolator units, exposure chambers, aerosol generators, real-time gas and aerosol monitors, analytical support, and the biomedical research items may require the use of several laboratory rooms. Inhalation studies that meet today's standards usually can only be performed successfully by a well trained, experienced, and adequately funded team of specialists working in a sophisticated laboratory.

V. Problems Encountered

The problems encountered in an inhalation study are largely practical ones, and they yield to the problem-solving methods of several key scientific disciplines (Table 1.3). Maintaining healthy animal subjects, a real problem at times, requires experts in the disciplines of microbiology, nutrition, physiology, and veterinary medicine. Establishing a well characterized study atmosphere is dealt with by rigorous application of the principles of engineering, physics, and chemistry. Monitoring animals during exposures can be a challenging problem in itself. It is solved using equipment, knowledge, and techniques from the discipline of physiology. Acquiring and processing tissues, especially lung tissue, also present several challenges; the disciplines of anatomy, biology, and physiology are required to meet them. Although the problems encountered in inhalation studies are significant and often unique, they are solvable by the application of sound scientific principles that have been established over the past four decades.

TABLE 1.3
Summary of Commonly Encountered Problems and the Scientific Disciplines that Permit Their Solution

Problems	Disciplines
Maintaining healthy subjects	Microbiology, nutrition, physiology, veterinary medicine
Air cleaning and conditioning	Engineering
Aerosol and gas generation	Physics, chemistry, engineering
Characterization of the study atmosphere	Physics, chemistry
Monitoring animals during exposure	Physiology, engineering
Acquiring and processing tissue samples	Anatomy, biology, physiology

VI. Scope of this Book

A. Overview

The basic requirements that essentially all inhalation studies have in common will be the focus of this book. It is not possible for a single volume to cover all types of studies, but the foundations can be treated here. Although studies with humans and laboratory animals have similar basic requirements, this book is largely limited to animal studies (with a focus on the rat). Animal studies are not only more common than human studies, but they also are usually more expensive and more demanding. Sound ethical principles require that well conducted animal studies precede studies with human subjects. If one can conduct a good animal inhalation study, similar human studies probably will be relatively easy, but not vice versa. Several useful reference works on inhalation toxicology can be found that provide essential guidance.[11-19]

B. Focus of the Chapters

Each chapter focuses on an important aspect of laboratory animal inhalation studies. The chapters are written to be independent, in that their order is immaterial, and each can largely stand alone. Because they are written by specialists in a variety of disciplines, their organization differs somewhat. Chapter 2, by William Mautz and Michael Kleinman, relates to the health and welfare of the animal subjects and to the process of obtaining approval of a animal research protocol from an institutional review committee. In a real sense, Chapter 2 is an ethics chapter, for it describes the prevention of disease, provisions for comfortable housing, and maximizing the study benefits while minimizing the stresses. Chapter 3, by Robert Phalen, is largely an engineering exercise. It covers the specialized area of cleaning and conditioning the "throughput" air to be used. Topics such as particle and gas scrubbing, temperature and humidity adjustment, and selection of equipment are presented. Chapter 4, by Michael Kleinman and Robert Phalen, is oriented toward physics and chemistry, as well as engineering. It deals with the complex specialties of generating study atmospheres and measuring their physical and chemical properties (i.e., atmosphere characterization). Chapter 5, by Robert Phalen, deals with exposure methods. Because such methods are so varied (exposures using chambers, manifolds, helmets, masks, canulas, etc., have been performed previously), the chapter focuses on nose-only exposures — perhaps the most versatile method of inhalation exposure. Chapter 6, by William Mautz, is based in physiology; it covers methods used to monitor animals during their exposure. Such monitoring is useful both to observe effects and to estimate the inhaled "dose" of the study material. Chapter 7, by Ronald Rasmussen, describes the anatomical methods used for acquiring and preparing tissue samples for further study. Euthanasia, fixation, embedding, sectioning, staining, and other demanding techniques are described. Chapter 8, by Deepak Bhalla, deals with two very important specialized sampling methods: bronchioalveolar lavage and blood collection.

The techniques described in this book look at the respiratory tract both as a potential target, and as a portal of entry to the body for inhaled materials. Taken together, the various chapters can provide a great deal of guidance to those who are beginning to perform inhalation studies or wish to modernize their ongoing studies.

References

1. Morrow, P. E., Bates, D. V., Fish, B. R., Hatch, T. F., and Mercer, T. T., Deposition and retention models for internal dosimetry of the human respiratory tract (report of the International Commission on Radiological Protection: ICRP: Task Group on Lung Dynamics), *Health Phys.*, 12,173, 1966.
2. Bair, W. J., Overview of ICRP respiratory tract model, *Radiat. Prot. Dosim.*, 38,147, 1991.
3. *ISO: Air Quality-Particle Size Fraction Definitions for Health Related Sampling*, ISO/TR 7708-1983 (E), International Standards Organization, Geneva, 1983.
4. Air Sampling Procedures Committee, American Conference of Governmental Industrial Hygienists (ACGIH): *Particle Size-Selective Sampling in the Workplace,* American Conference of Governmental Industrial Hygienists, Cincinnati, 1985.
5. Cuddihy, R. G. and Yeh, H. C., Respiratory tract clearance of particles and substances dissociated from particles, in *Inhalation Toxicology: The Design and Interpretation of Inhalation Studies and Their Use in Risk Assessment,* Dungworth, D., Kimmerle, G., Lewkowski, J., McClellan, R., and Stöber, W., Eds., Springer-Verlag, New York, 1988, chap. 11.
6. Weibel, E. R., *Morphometry of the Human Lung,* Academic Press, New York, 1963.

7. Guilmette, R. A. and Boecker, B. B., Eds, *Respiratory Tract Dosimetry: Proceedings of a Workshop held in Albuquerque July 1–3, 1990,* Ashford Kent: Nuclear Technology Publishing, 1991 (also, *Radiat. Prot. Dosim.,* 38(1–3), 1991).
8. Yeh, H. C. and Schum, G. M., Models of human lung airways and their application to inhaled particle deposition, *Bull. Math. Biol.*, 42,461, 1980.
9. Phalen, R. F., Yeh, H. C., and Prasad, S. B., Morphology of the respiratory tract, in *Concepts in Inhalation Toxicology,* Second ed., McClellan, R. O. and Henderson, R. F., Eds., Taylor and Francis, Washington, D.C., 1995, chap. 5.
10. Yeh, H. C., Schum, G. M., and Duggan, M. T., Anatomic models of the tracheobronchial and pulmonary regions of the rat, *Anat. Rec.*, 195, 483, 1979.
11. Willeke, K., Ed, *Generation of Aerosols and Facilities for Exposure Experiments,* Ann Arbor Science, Ann Arbor, MI, 1980.
12. Leong, B. K. J., Ed., *Proceedings of the Inhalation Toxicology and Technology Symposium*, Ann Arbor Science, Kalamazoo, MI, 1981.
13. Phalen, R. F., *Inhalation Studies: Foundations and Techniques,* CRC Press, Boca Raton, FL, 1984.
14. Salem, H., *Inhalation Toxicology: Research Methods, Applications and Evaluation,* Marcel-Dekker, New York, 1987.
15. Dungworth, D., Kimmerle, G., Lewkowski, J., McClellan, R., and Stöber, W., Eds., *Inhalation Toxicology: The Design and Intrepretation of Inhalation Studies and Their Use in Risk Assessment*, Springer-Verlag, New York, 1988.
16. McClellan, R. O. and Henderson, R. F., Eds., *Concepts in Inhalation Toxicology*, Hemisphere Publishing, New York, 1989.
17. McClellan, R. O. and Henderson, R. F., Eds., *Concepts in Inhalation Toxicology*, Second ed., Taylor and Francis, Washington, D.C., 1995.
18. Willeke, K. and Baron, P. A., Eds, *Aerosol Measurement: Principles, Techniques and Applications*, Van Nostrand Reinhold, New York, 1993.
19. Jenkins, P. G., Kayser, D., Muhle, H., Rosner, G., and Smith, E. M., Eds, *Respiratory Toxicology and Risk Assessment: Proceedings of an International Symposium*, Wissenschaftliche Verlagsgesellschaft mbH, Stuttgart, Germany, 1994.

Chapter 2

Animal Quality and Protocol Approval

William J. Mautz and Michael T. Kleinman

Contents

0-8493-3343-1/97/$0.00+$.50

I. Objectives

Two primary requirements in using animal models in scientific research are that the experimental results be reproducible and that the animal subjects be treated humanely. Experimental protocols are designed to reduce sources of variation that affect the parameters under study so that the experimental outcomes can be interpreted as the direct result of a test condition. The second requirement, ethical treatment of animals, is inextricably bound to the first, for stressful conditions introduce physiological changes that can affect the outcome and interpretation of experimental results. Animal subjects must be maintained under clean and well controlled living conditions, and, for most studies, the animals must be free of confounding disease. For inhalation exposure studies, it is particularly important to ensure freedom from respiratory tract diseases.

The main purpose of this chapter is to describe the basic requirements for quality control of animals in inhalation exposure studies. A secondary purpose is to provide guidance on obtaining protocol approval from an institutional oversight committee.

II. Scope and Limitations

Inhalation studies have involved a large variety of mammalian animals, including rodents, lagomorphs (rabbits), carnivores, primates, and ungulates, as well as nonmammalian vertebrates. Different animal species have widely varying requirements for housing and maintenance, and there have been detailed reviews on their maintenance requirements.[1-3] The National Research Council's Animal Care Panel has recently revised the *Guide for the Care and Use of Laboratory Animals*.[4] This guide, which is the basis for facility accreditation, covers: institutional policies, housing and management, veterinary care, and physical facilities. The 1996 edition (currently in press) should be available from the National Academy Press, 2101 Constitution Ave., NW, Box 285, Washington, D.C. 20055. Laboratory rats and mice are the most common animals used in inhalation studies, and this chapter will focus on quality control issues concerning rodents for inhalation exposure studies. An excellent journal covering animal care and use is *Lab Animal*, which may be obtained from: Lab Animal, P.O. Box 5054, Brentwood, TN 37024-5054.

III. Problems Encountered

A. Obtaining Disease-Free Animals

The most difficult problem in animal quality control for inhalation studies is to ensure that animals are free of respiratory disease. Respiratory tract infections are very common in rodents, and latent or mild infections may not necessarily result in mortality or outward signs of infection. However, lung tissues examined microscopically can show the pathology of pneumonitis, and these features may seriously confound the interpretation of the effects of experimental exposures. Furthermore, inhalation exposures can make animals more susceptible to acquiring or expressing respiratory tract infections, and disease effects may be much more prominent in exposed groups compared to purified air control groups.

Respiratory tract infections are often highly contagious. Animal suppliers have difficulty supplying disease-free animals, and animals housed in the laboratory can acquire diseases from other laboratory animals, from wild rodents if they have access to the facility, or from people who have had contact with other diseased rodents.

B. Maintaining Animals Disease Free

Maintenance of quality animal care requires diligent oversight. Animal care must be performed by a well trained staff. Housing, procedures for care, and experimental procedures must be approved, documented, and periodically reviewed by an oversight committee that is staffed with people who are experts in animal care and scientific investigation.

IV. Animal Study Considerations

A. Quality Control Analyses of Animals

Prevention of and monitoring for respiratory tract infections that could compromise an inhalation study are crucial aspects of quality control. Animals should be obtained from a reliable supplier that offers certified pathogen-free stocks of animals. These animals are raised under barrier conditions to isolate them from disease sources, and they undergo periodic testing to determine if they have been exposed to diseases. The introduction of diseases into the suppliers' colonies is a constant threat, and it is possible to receive a contaminated shipment in which disease was newly acquired at the supplier facility or during shipment; therefore, the investigator should make independent analyses from a sample of animals in a shipment upon their arrival. This sample is usually 10 to 15 animals for shipments of between 100 and 500 rodents. The analyses should include a blood test for a panel of rodent bacterial and viral pathogens (Table 2.1). These analyses can be performed by commercial laboratories, and they usually have a rapid turnaround time. In rats, a viral pathogen, Sendai (parainfluenza), and a mycoplasma pathogen are the major infectious agents that pose the most serious threat to pulmonary health.[5] Quality control analyses also should include gross and histopathologic examinations of the lungs. Examination of lung tissues can reveal the presence of

TABLE 2.1
Common Rodent Respiratory Tract Pathogens[5]

Viral pathogens
Parainfluenza (Sendai, types 2,3)
Pneumovirus (respiratory syncytial virus PVM)
Bacterial pathogens
Streptococcus pneumoniae
Staphylococcus aureus
Corynebacterium sp.
Pasturella sp.
Bordetella sp.
Hemophilus sp.
Klebsiella pneumoniae
Mycoplasma sp.

perivascular infiltrates of inflammatory cells, peribronchial lymphocytic follicles, and foci of interstitial pneumonia. Long-term chronic studies should employ a sentinel population of animals periodically analyzed for presence of infectious disease. Examination of animal condition may reveal symptoms of infectious disease; animals should be observed for abnormal grooming behavior, failure to eat or drink, and evidence of discharges from the nose or eyes (Table 2.2).

B. Purified Air for Animal Housing

Animal housing for inhalation studies has a unique requirement of a purified air supply. Most institutional animal quarters have effective temperature and humidity control (20 to 23°C and greater than 20% relative humidity, for common laboratory rodents), but they have inadequate control over airborne contaminants for inhalation toxicology research. Air should be filtered for both particles and trace contaminant gases. The air supply to an animal room should have about 15 to 20 air changes per hour, and recirculated air should be limited to less than 75% of total ventilation. The filtration system should be capable of removing outdoor air pollutants such as ozone, sulfur dioxide, and nitrogen oxides and indoor pollutants such as ammonia and hydrocarbons. Important particulate air pollutants that should be removed are animal dander, dust, viruses, and bacteria.

Contaminant gases can be removed by chemical filtration using charcoal and buffered potassium permanganate-impregnated on activated alumina pellets (Purafil, Inc.; Doraville, GA). Particles are removed in two stages of filtration. A coarse particle filter will remove large particles and prolong the life of the more expensive fine particle filter downstream. A high efficiency

TABLE 2.2
Symptom Categories and Quantification (on a Scale of 0 to 3) of Pain and Distress in Animals[a]

Body Weight	
0	Normal
1	<10% weight loss
2	10–20% weight loss
3	>20% weight loss
Physical appearance	
0	Normal
1	Lack of grooming
2	Rough coat, shedding, nasal and/or ocular discharge
3	Very rough coat, abnormal posture, enlarged pupils
Measurable clinical signs	
0	Normal
1	Small changes of potential significance
2	Body temp. change 1–2°C, cardiac/respiratory rates increased up to 30%
3	Body temp. change >2°C, cardiac/respiratory rates increased up to 50%, or markedly reduced
Behavior	
0	Normal
1	Minor changes, decreased appetite
2	Abnormal, reduced mobility, decreased alertness, inactivity
3	Unsolicited vocalizations, self-mutilation, either very restless or immobile
Responses to stimuli	
0	Normal
1	Minor depression or exaggeration of response
2	Moderately abnormal responses
3	Violent reactions or comatose state

[a] Adapted from Olfert, E. D., *Lab Animal,* 25(3), 38, 1996.

particulate air (HEPA) filter unit can then remove nearly 100% of particles of all sizes, with the minimum efficiency of 99.97% for 0.3 μm-diameter particles. Chapter 3 covers the principles and techniques of air purification in detail.

Important sources of airborne contaminants and pathogens are the animal caging system and animals themselves. An air barrier caging system (Figure 2.1) with both chemical and HEPA filtration will remove gases and particles from the room air and supply freshly purified air flowing though the rodent cages. The system shown uses horizontal laminar air flow to prevent air from passing directly between animal cages and to inhibit the spread of airborne infections. Ammonia, from decomposition of urine and from bacteria in feces, and dander particles from the animals are blown directly into the

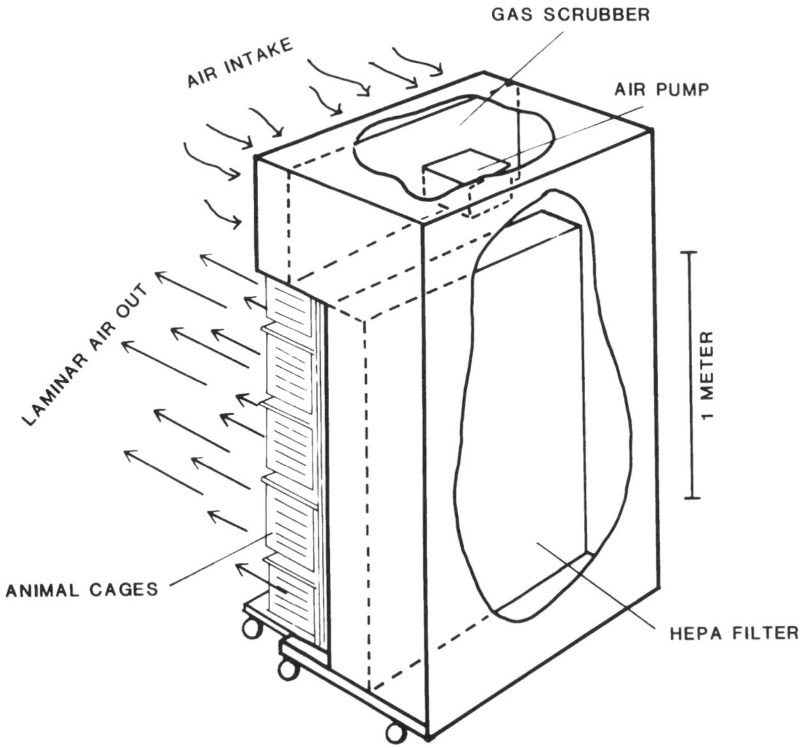

Figure 2.1
Laminar-flow purified air barrier unit used for rodent cage rack. (From Phalen, R. F., *Inhalation Studies: Foundations and Techniques,* CRC Press, Boca Raton, FL, 1984, 199.)

room and filtered before the room air is resupplied to the animals. An additional control on ammonia generation is described below.

C. Caging for Animals

Laboratory rodents are held in cages with standardized minimum space requirements (Table 2.3). Rodents should have continuous access to food and water, which is usually supplied in individual food hoppers and water bottles; water may also be supplied via an automatic lick-controlled watering system. The supplies in food hoppers and bottles can be visually inspected. If an automatic watering system is used, individual lines supplying each cage should be checked daily to ensure they are not clogged or leaking into the cages, and the water filter should be changed monthly. Cages may be constructed of wire, sheet metal, or plastic. Galvanized wire caging can give rodents excessive levels of zinc if they chew on the wire, while polished stainless steel wire caging is less readily chewed by rodents. Plastic bins require a bedding material for absorption of waste. Bedding may be made of wood chips, paper products,

TABLE 2.3
Space Recommendations for Laboratory Rodent Cages

Animals	Body weight (g)	Floor area per animal (cm^2)	Cage height (cm)
Mice	<10	39	12.7
	10–15	52	12.7
	16–25	77	12.7
	>25	97	12.7
Rats	<100	110	17.8
	100–200	148	17.8
	201–300	187	17.8
	301–400	258	17.8
	401–500	387	17.8
	>500	>452	17.8
Hamsters	<60	65	15.2
	60–80	84	15.2
	81–100	103	15.2
	>100	123	15.2
Guinea pigs	to 350	377	17.8
	>350	652	17.8

Source: National Research Council, Animal Care Panel, *Guide for the Care and Use of Laboratory Animals,* DHEW Publ. No. (NIH) 86-23, U.S. Government Printing Office, Washington, D.C. 20402, 1095.

corn, or other cellulose material. These materials are invariably dusty, and air filtration systems cannot protect animals living in close contact with bedding from inhaling substantial quantities of particles. The inhaled particle load may induce a background level of pulmonary inflammatory response that may substantially interfere with the evaluation following an inhalation exposure.

The use of wire-bottomed cages will avoid the need for organic bedding compounds. Feces and urine are collected in trays below the wire floors. These are still a source of ammonia, as excreta are decomposed by bacteria. Addition of a layer of rock salt to the collection trays will dry the feces and urine and suppress the production of ammonia. Salt is also dust free (due to an aqueous surface film that forms at normal relative humidity), and relatively inexpensive in bulk quantities. A disadvantage is that salt is heavy, and regular maintenance presents a large mass of salt for disposal.

D. Maintaining Clean Housing

Animal facilities should be kept scrupulously clean, all supplies should be well organized, and the facility should be free of clutter. The rooms should be tightly closed to prevent entry of insects and wild rodents. Flooring should be well drained and easily cleaned, and countertops and sinks should be orderly and easily disinfected. Housing areas and daily care should be maintained by qualified animal technicians who perform regular cleaning and disinfection procedures, as well as recordkeeping (see Section G below). Separate rooms must be used for housing each animal species, for cage and dish washing, for food preparation and storage, and for storage of clean caging. Additional areas will be needed for storage of cleaning supplies, vivarium work attire (laboratory coats, masks, gloves, boots, and hair covers), carts, and drugs. A checklist is valuable as a reference for regular cleaning and disinfection maintenance (Table 2.4).

TABLE 2.4
Sample Checklist for Animal Facility Maintenance

- Check general health of every animal daily. Check food and water. Flush automatic watering lines.
- Record the temperature and relative humidity in the animal rooms daily.
- Mop animal facility floors daily using a biocidal solution.
- Wear clean gloves when handling animals.
- Change cage pans or bedding two times per week.
- Wash cages and cage racks once per week. Cover automatic water-system nipples with plastic shield before washing. Use thermal tags on cage washer loads to verify washing temperature. Clean cages when they are brought to the cage washer.
- Wash trash cans weekly.
- Clean countertops daily.
- Clean walls, isolator (if used), and light fixtures monthly.
- Keep clean (fresh washed) equipment storage time short.
- Change pre-filters in the isolator units, and check on chemical scrubber condition annually.
- For automatic water systems: flush system daily and change filter monthly.

E. Fumigation

An infectious disease that gains entry to an animal facility can be very difficult to eradicate. Despite diligent attention to daily cleaning and disinfecting procedures, disease-producing organisms may survive in the facility in inactive forms and subsequently infect new batches of animals. Chronic exposure studies are particularly vulnerable, because individual animals are maintained for extended periods, perhaps for the lifetime of the animals.

When disease is discovered, the facility should be fumigated with a gaseous disinfectant, or fogged with a liquid disinfectant droplet aerosol. Fogging disinfection of a 100-m^3 volume room can be achieved with a 1-*l* solution of Clidox-S (Pharmacal Research Laboratories; Nangatuck, CT). Fumigation with gas, such as formaldehyde vapor, will better penetrate to recessed surfaces than will liquid aerosol. Paraformaldehyde is a solid polymer that decomposes to formaldehyde at 232°C. Fumigation can be accomplished safely by using a timer switch and hot plate (formaldehyde is extremely hazardous in high concentrations). An aluminum or stainless steel pan containing 400 g paraformaldehyde (for a 100-m^3 room) is set on the hot plate. An electric frying pan with a calibrated thermostat may be substituted for the hot plate and pan. With the air conditioning systems turned off, formaldehyde is vaporized over a period of several hours as the hot plate is automatically turned on and off. The following day, the air conditioning system is turned on by personnel wearing appropriate respirators, and, after several hours, the rooms can again be entered. If fumigated areas are not ventilated promptly and thoroughly, the area may be uninhabitable for several days while formaldehyde is out-gassing from building surfaces.

F. Protocol Approval

An institutional oversight committee is responsible for ensuring that laboratory animals are properly housed and cared for and that all laboratory procedures are both necessary and humane. In order to accomplish this goal, the committee must review and approve a protocol documenting the procedures to be performed, the facilities to be used, the personnel to be involved, and the procedures to be used for animal care. The committee should be composed of a veterinarian and other experts knowledgeable in the scientific research purpose and the procedures used in animal care. However, there may be lay people representing the local community on such committees, and protocol language should be direct and free of overly technical jargon.

Advance detailed planning is necessary to determine the numbers of animals that will be needed, as well as all the procedures that will be performed with laboratory animals. The basic requirements of a protocol submission include a listing of the personnel involved in handling animals and their backgrounds and experience with the species of laboratory animal used in the study. The animal housing facility and all laboratories where live animals will be present must be listed, so that periodic inspections can be conducted by the committee. A listing of species and numbers of animals to be used must be provided, and the procedures are secondarily classified by U.S. Department of Agriculture (USDA) pain categories, which require the use of different levels of anesthetics or analgesics. A listing of all chemical agents (medications, anesthetics, and study materials) to be used is required and should include the dose, route of administration, and the rationale. If animals are going to be bred, then additional information about breeding facilities, along with the numbers and fates of young produced and the rationale for breeding, is necessary.

In addition to the basic summary described above, the protocol document will have a description of the purpose, importance, and overall procedures involved in the research project. This narrative must clearly state the rationale for the study in terms of potential benefits of the project for humans, animals, or society; detailed descriptions of the procedures to be performed (including the numbers of animals to be used in each procedure); and methods used to ensure that alternatives to any proposed painful procedures with animals were considered and reviewed. If surgery is required, then additional documentation of the protocol and a plan for postoperative care (and how this care will be documented) must also be described. Finally, the method of assessing pain, distress, and discomfort of animals, how these conditions will be managed, and the method of euthanasia should be documented.

G. Recordkeeping

A documented record of animal care procedures is necessary for oversight of laboratory animal studies. Approved protocols should be posted in each room where the procedures are used. Animals are identified by means of approved cage cards, ear tags, tattoos, or other appropriate methods. Information associated with the cage cards should include the species and strain, source of the animal, the name and location of the responsible investigator, and date of receipt of the animal(s). The eventual disposition of each animal also should be recorded. In the animal quarters, a record of ambient conditions should be made either with recording hygrothermographs or by recording the daily maximum and minimum temperature and a daily measure of relative humidity.

V. Suppliers of Equipment

A. Animal Suppliers (Rodent)

Charles River Laboratories, Inc.
251 Ballardvale Street
Wilmington, MA 01887

Harlan Sprague-Dawley, Inc.
P.O. Box 29176
Indianapolis, IN 46229

Hilltop Lab Animals, Inc.
P.O. Box 183
Scottdale, PA 15683

Simonsen Laboratories, Inc.
1180-C Day Road
Gilroy, CA 95020

B. Rodent Caging and Barrier Filtration Systems and Media

Lab Products, Inc.
255 W. Spring Valley Avenue
P.O. Box 808
Maywood, NJ 07607

AirTech
2007 Raymer Street, Suite S
Fullerton, CA 92633

Wheelabrator/Westates Carbon, Inc.
2130 Leo Avenue
Los Angeles, CA 90040

Purafil, Inc.
2654 Weaver Way
Doraville, GA 30340

C. Disinfectants

Pharmacal Research Laboratories
33 Great Hill Road
Nangatuck, CT 06770

Sigma Chemical Co.
P.O. Box 14508
St. Louis, MO 63178

Fisher Scientific
711 Forbes Avenue
Pittsburgh, PA 15219

D. Veterinary Diagnostic Laboratories

Microbiological Associates
Life Science Center
9900 Blackwell Road
Rockville, MD 20850

USC Diagnostic and Research Support Laboratory
2250 Alcazar Street, CSC 127
Los Angeles, CA 90033

Research Animal Diagnostic and Investigative Laboratory
1600 East Rollings
Columbia, MO 65211

References

1. Baker, H. J., Lindsey, J. R., and Weisbroth, S. H., Housing to control research variables, in *The Laboratory Rat,* Vol. 1, Baker, H. J., Lindsey, J.R., and Wesibroth, S. H., Eds. Academic Press, New York, 1979, chap. 8.
2. McPherson, C. W., Laws, regulations, and policies affecting the use of laboratory animals, in *Laboratory Animal Medicine,* Fox, J. G., Cohen, B. J., and Loew, F. M., Eds., Academic Press, New York, 1984, chap. 2.
3. National Research Council, Animal Care Panel, *Guide for the Care and Use of Laboratory Animals,* DHEW Publ. No. (NIH) 86-23, Superintendent of Documents, U.S. Government Printing Office, Washington, D.C. 20402, 1985 (revision due in 1996).
4. Shalev, M., The 1996 guide, *Lab Animal,* 25(3), 16, 1996.

5. Castleman, W. L., Spontaneous pulmonary infections in animals affecting studies on comparative lung morphology and function, *Am. Rev. Resp. Dis.*, 128(Suppl.), S83, 1983.
6. Olfert, E. D., Considerations for defining an acceptable endpoint in toxicological experiments, *Lab Animal,* 25(3), 38, 1996.
7. Phalen, R. F., *Inhalation Studies: Foundations and Techniques*, CRC Press, Boca Raton, FL, 1984.

Chapter 3

Cleaning the Air

Robert F. Phalen

Contents

0-8493-3343-1/97/$0.00+$.50

I. Introduction

A. Objectives

Inhalation studies must meet rigorous standards with respect to the generation and control of the exposure environment. Purified air, having precisely adjusted temperature and humidity, is intentionally polluted by adding substances having known and stable compositions, concentrations, and physical characteristics. No unwanted contaminants should be present. In most studies, a matched control population of subjects is exposed either simultaneously (preferred) or serially (not preferred) to purified air (with matched temperature and humidity) which contains insignificant levels (at most) of the pollutants. One requirement of such studies is the ability to clean and condition large quantities of impure, contaminated ambient air. To meet this requirement, a suitable air purification and conditioning system must be designed, installed, tested, and operated. The objective of this chapter is to present the fundamental issues, along with the problems and solutions, involved in providing enough purified air at the proper humidity and temperature to support an inhalation study. A more complete treatment of air cleaning for inhalation studies has been published.[1] Much of what follows is from this reference.

B. Commonly Encountered Air Contaminants

The air used in inhalation studies is ultimately derived from ambient air; rarely is the purity of this air sufficient to use in an exposure without prior cleaning. All of the significant indoor and/or outdoor air pollutants must, therefore, be substantially removed from the air prior to its use. These pollutants include natural and anthropogenic particles, as well as gases and vapors (inorganic, organic, and mixtures). Additionally, the ambient air can have a temperature and humidity that are inappropriate and thus require adjustment.

The devices used for the filtering and conditioning of ambient air will offer resistance to the flow of air and require an air compressor in order to provide the needed air flow rate. Air from a compressor is usually stored in a high-pressure holding tank before being piped into the laboratory. Typically, compressed air is quite dirty, containing a large variety of undesirable contaminants (Table 3.1). Aside from atmospheric dust and other particles, compressed air is likely to contain aerosolized lubricants, unburned petroleum fuel, fuel combustion products, water droplets and vapor, and scales from internal surfaces of tanks, pipes, and fittings.[2] The use of filters and oil-less compressors will eliminate many contaminants; however, additional air purification is almost always necessary.

TABLE 3.1
Commonly Encountered Airborne Contaminants in Unfiltered Compressed Air

Source	Contaminants
Ambient air	Soil and other dusts, miscellaneous gases and vapors
Compressor lubrication	Oil mist, oil decomposition products
Fuel, fuel combustion	Hydrocarbons, carbon monoxide, carbon dioxide, nitrogen dioxide, aldehydes, water vapor, acid-forming gases
Holding tank, pipes, and other surfaces	Metal, metal oxide particles, sediment particles

Source: Phalen, R. F., *Inhalation Studies: Foundations and Techniques,* CRC Press, Boca Raton, FL, 1984, 78.

II. Gas and Vapor Removal

In addition to the gases and vapors that contaminate supply air, water vapor often will have to be removed to prevent condensation in the air supply system. Table 3.2 gives the amount of water vapor in saturated air as a function of temperature. Relative humidity is defined as the percentage of saturation with water vapor at the prevailing temperature. The major methods for removal of water vapor from a gas stream are adsorption, absorption, and condensation, usually stimulated by cooling. Each method has its own removal efficiency, capacity, and cost. (Removal efficiency as used here is the number of milligrams of water remaining per liter of air after the removal process has reached equilibrium.) An example of an efficient drying method is extreme cooling. Cooling to –70°C with dry ice and acetone leaves about 0.01 mg of water per liter of air at equilibrium (i.e., 100% relative humidity at this low temperature). This moisture content at room temperature represents a relative humidity of only about 5%. Ordinary refrigerated-air cooling units can be used to reduce the relative humidity of the air to about 20 to 30%, upon re-warming to room temperature.

Cooling of an air stream, as seen in Table 3.2, reduces the maximum amount of water vapor that is carried in the air at saturation. As an example, consider a refrigeration unit that cools an air stream to 30°F and is followed by a heater that increases the air stream temperature to 70°F before passage into an exposure chamber. From Table 3.2, the cooling will result in a maximum water content of 4.5 g/m^3 of air. Upon warming to 70°F, this 4.5 g/m^3 represents about 23% of the saturation value of 18.5 g, resulting in a relative humidity of about 23% in the chamber.

Removal of water vapor by absorption or adsorption is achieved by solid or liquid desiccants. Such desiccants may have an added indicator that changes color when the material is water saturated, as well as the potential for regeneration by heating or flushing with dry air. Some disadvantages of desiccants

TABLE 3.2
Amounts of Water Vapor in Clean Air at Saturation (100% Relative Humidity) as a Function of Temperature

Air temperature (T)		Grams of water per m³ of air (at temperature T)
°F	°C	
0	–17.8	1.23
10	–12.2	1.86
20	–6.7	2.94
30	–1.1	4.49
40	4.4	6.55
50	10.0	9.40
60	15.6	13.27
70	21.1	18.45
80	26.7	25.31
90	32.2	34.23
100	37.8	45.74
110	43.3	60.39
120	48.9	78.86
130	54.4	101.8
140	60.0	130.2
150	65.6	165.1
160	71.1	207.3
170	76.7	258.1
180	82.2	319.0
190	87.8	391.2
200	93.3	476.2

Source: Phalen, R. F., *Inhalation Studies: Foundations and Techniques*, CRC Press, Boca Raton, FL, 1984, 78.

include relatively long equilibration times and the buildup of heat during the desiccation process. Solid desiccants include calcium sulfate, activated alumina, and molecular sieves (Table 3.3). Liquid desiccants include sulfuric acid, triethylene glycol, sodium hydroxide, potassium hydroxide, calcium chloride, and glycerol (Table 3.4). Nelson[3] has a useful discussion of desiccants that may be consulted by designers.

In addition to the particulate pollutants in compressed air, ambient pollutant gases may also be present in sufficient concentrations to require removal. Such gases include ozone, sulfur dioxide, and oxides of nitrogen. Removal of such gases can be performed by adsorption onto a large surface area (activated charcoal, for example) or by chemical oxidation to produce a product that is easily removed. For this latter purpose, a catalyst bed or an oxidizer bed can be used to promote reactions such as:

TABLE 3.3
Efficiencies in Terms of Residual Water Content of Selected Solid Desiccants Used for Drying 25°C Nitrogen Gas at 225 cc/min Through a Bed 14 × 450 mm

Desiccant	Average efficiency (mg/*l* of air)	Regeneration time (hr), temp. (°C)
Calcium sulfate (Drierite®)	0.067	1–2, 200–225
Silica gel	0.070	12, 118–127
Activated alumina	0.0029	6–8, 175–400
Anhydrous magnesium perchlorate	0.0002	48, 245
Molecular sieve 5A (Union Carbide®)	0.0039	—

Source: Phalen, R. F., *Inhalation Studies: Foundations and Techniques,* CRC Press, Boca Raton, FL, 79.

TABLE 3.4
Properties of Selected Liquid Desiccants

Desiccant	RH achieved at 21°C (%)	Solution conc. (%)	Operating temperature range (°C)
Sulfuric acid	5–20	60–70	21–49
Triethylene glycol	5–10	70–95	16–43
Sodium hydroxide and potassium hydroxide	10–20	Saturated	29–49
Calcium chloride	20–25	40–50	32–49
Glycerol	30–40	70–80	21–38

Note: The relative humidity (RH) achieved is the range of values measured over the saturated aqueous solution at 21°C.

Source: Phalen, R. F., *Inhalation Studies: Foundations and Techniques,* CRC Press, Boca Raton, FL, 1984, 79.

$$2O_3 \rightarrow 3\ O_2$$

$$CO + 1/2\ O_2 \rightarrow CO_2$$

$$NO + 1/2\ O_2 \rightarrow NO_2$$

$$SO_2 + 1/2\ O_2 \rightarrow SO_3$$

$$SO_3 + H_2O \rightarrow H_2\ SO_4 \text{ (particles)}$$

Adsorbants that trap pollutant gases include activated carbon, molecular sieves, and activated alumina. Oxidation can be accomplished by such means as a platinum catalyst bed operated at elevated temperature or a potassium permanganate bed. Potassium permanganate, an excellent oxidizer for many pollutant gases, is available impregnated on activated alumina pellets (Purafil®,

Purafil, Inc.; Doraville, GA), which are composed of a mixture of oxides of copper and manganese.

Removal of organic vapors, other than low molecular weight species, can be accomplished by two means: adsorption onto an agent such as activated charcoal or high-temperature oxidation. One method, attributed to Kushnetz et al. by Nelson,[3] is passage through a 5-cm diameter tube filled with copper shavings maintained at 1250°F. This method requires cooling of the exit gas stream and removal of oxidation products by passage through an acid bubbler, a filter coated with sodium hydroxide, a charcoal bed, and a particle filter.

III. Particle Removal

Removal of particles from a gas stream can be achieved by electrostatic precipitation, thermal precipitation, or filtration. Precipitation methods will not be discussed, as they are seldom used in the laboratory for cleaning air. Air for laboratory use usually is mechanically filtered because of the potential of this method for efficient removal of even the smallest particles and the simplicity of operation.[4] Membrane filters can be efficient in removing particles but often have the undesirable feature of relatively high resistance to air flow (i.e., pressure drop). Fiber filters generally have much lower resistance, and these types of filters are available in forms that have very high capture efficiency for particles of all sizes. One disadvantage of high-efficiency filters is their tendency to "blind" or clog as they become filled with captured particles. This problem is usually dealt with by addition of a lower efficiency (coarse) filter upstream, by pleating the filter to increase its overall surface area, and by continuous measurement of the pressure drop across the filter as an indicator of the need for replacement. High-efficiency filters usually cannot be cleaned, so they are discarded when their pressure drop becomes too high. An excellent practical reference that covers many important design considerations for particle filtering systems is the *Nuclear Air Cleaning Handbook* by Burchsted, Kahn, and Fuller.[5] Much of the following is adopted from this reference. In general, the strict air-filtering requirements established for nuclear facilities also should be met in the inhalation toxicology laboratory.

Final filtering of air is often done with HEPA (high-efficiency particulate air)-type filters. This type of filter (formerly called AEC, CWS, and absolute filter) is a disposable, deep-pleated, dry-type fiber filter in a rigid wood, plastic, or metal case. These filters have very fine inorganic fibers in a matrix of larger fibers. As mentioned in Chapter 2, HEPA filters have very high capture efficiencies, typically 99.9% or more, for removal of submicrometer particles. This efficiency is due to the fact that larger particles are effectively deposited by sedimentation and by interception (i.e., direct contact with filter fibers), while small particles are pushed into contact with the filter fibers by diffusion. HEPA filters are commercially available in a variety of sizes and shapes (see Section VII of this chapter).

Because HEPA filters can clog rapidly when presented with coarse particles and because their replacement cost is relatively high, they are installed behind a relatively inexpensive prefilter. Typically, prefilters include low-efficiency panel filters consisting of oil-coated or adhesive-coated assemblies of coarse fibers (very inexpensive types are available) and moderate and high-efficiency pleated or bag-type dry fiber filters. Low-efficiency filters, including furnace-type units found in the home, are very inefficient in capturing particles and are of limited use for protecting HEPA filters. Medium and high-efficiency filters usually have excellent efficiencies for particles a few micrometers or larger in diameter and are recommended. Selection criteria for prefilters should include their fire resistance and corrosive chemical resistance, as well as their particle capture efficiencies and replacement costs.

IV. Safety Considerations

From the point of view of safety, air cleaning and conditioning systems present a variety of problems. Powerful motors, fans, heavy components, hot surfaces, caustic chemicals, flammable substances, noise, vibration, and heavy electrical usage are some of the factors that must be considered. One appreciates the payoff of good safety practices often only when they are not in place and a serious injury occurs in the laboratory. The following discussion is only a starting point for safety and should not be considered to be all-inclusive.

Adequate space must be provided so that equipment can be worked with and maintained easily and does not get in the way of normal movement of people. Similarly, equipment should be placed in locations where accidental contact is unlikely. When it is impossible to keep hazardous equipment out of the work area, guards (for belts, fans, and switches) and pads (for sharp or hard objects) must be added, along with attention-getting warning labels and signs. An often overlooked aspect of safety is basic neatness. Cluttered, dirty, wet, or oily surfaces can greatly increase the probability of a serious accident. Another often overlooked hazard is noise, which may be stress inducing, for both people and the animal subjects. Excessive noise can permanently degrade hearing and impair concentration and communication. Noise is controlled by selecting quiet equipment, placing noisy items outside of the lab, reducing vibration, using good mounting systems, and applying sound-deadening pads and panels. In addition to these safety considerations, one must consider fire and earthquake safety. The laboratory must be made fire-resistant, and several appropriate extinguishers should be strategically placed. Heavy items must be secured to prevent their falling or tipping over in a large earthquake.

The management aspects of safety should not be overlooked. Personnel must know that safety is an important aspect of their jobs, and explicit safety training should be provided. In addition, safety inspections that include laboratory representatives and discussions about safety at lab meetings should occur regularly.

V. Problems Encountered

Most air pollutants are easy to remove; however, some are not. Nonreactive gases, such as carbon monoxide, may require the use of high-temperature or high-pressure catalyst or adsorbent beds to accelerate oxidation and removal. Contaminants generated by the air-cleaning system itself can be frustrating; some adsorbants are dusty, for example. Reliability is always a problem, especially when equipment and maintenance costs are minimized. Getting adequate space set aside for an air purification system is often difficult. Additional problems are related to maintenance and repairs, especially during an ongoing study when down time is harmful to the research. Potential problems are best dealt with before they occur. In other words, plan ahead, involve experts, and do not sacrifice quality and reliability for cost.

VI. Air Purification and Conditioning Systems

The system configuration for purifying and conditioning air will depend upon several factors, including types and concentrations of initial air contaminants, desired purity, required purified air flow rates and pressures, and acceptable ranges of humidity and temperature of the air. The following two designs, used at the Air Pollution Health Effects Laboratory at the University of California, Irvine, are examples of systems specifically designed for inhalation studies.

A. A Low-Pressure System

The system shown in Figure 3.1 was designed in 1974 by an engineer, William Dennison, to meet the criteria listed in Table 3.5. The system cost at the time, excluding design, was approximately $30,000 installed. This system has provided purified air reliability for more than 20 years.

Ambient air enters this system through a 10-cm diameter pipe from two selectable sources: either from above the laboratory roof or from indoors. Ambient air is first filtered for removal of coarse particles using a disposable, medium-efficiency dry pleated filter. Filtered air is passed through two parallel columns of permanganate-impregnated alumina pellets (Purafil®, Purafil, Inc.; Doraville, GA) 60 cm deep and 30 cm in diameter. The moisture in the inlet gas stream is adsorbed onto the alumina, dissolving the permanganate, which can then oxidize contaminants. This material removes NO, NO_2, NH_3, SO_2, O_3, most hydrocarbons, and some CO from the air. The Purafil® columns are followed by another medium-efficiency pleated filter for removal of potassium permanganate dust. Air exits the filter holder through a 10-cm diameter stainless steel pipe and enters an air temperature-humidity conditioning unit (Bemco PTHS Air Servo, Bemco, Inc.; Pacoima, CA). This unit has a refrigeration system

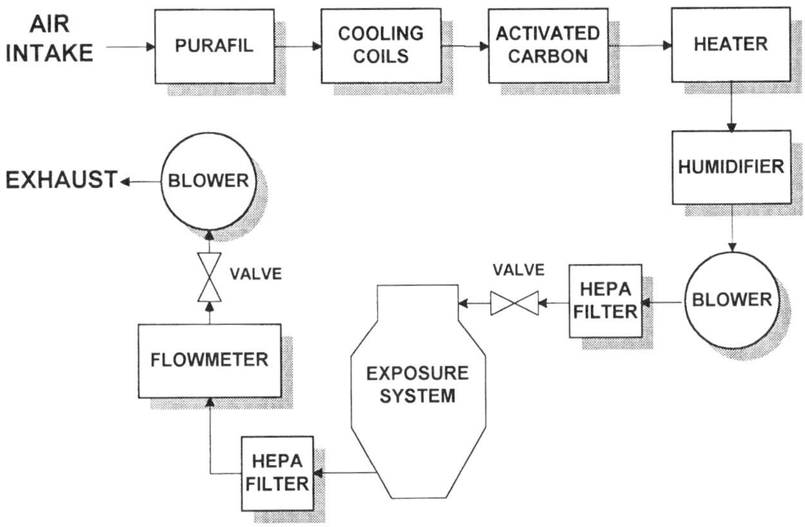

Figure 3.1
Low-pressure air purification system used to supply four Rochester-type, 1-m^3 volume exposure chambers.

TABLE 3.5
Design Parameters for an Air Purification System Used to Supply Air to Four, 1-m^3 Volume, Rochester-Type Exposure Chambers

Contaminant	Probable maximum inlet concentration (ppm)	Desired outlet concentration (ppm)
Nitric oxide	1	0.01
Nitrogen dioxide	1	0.01
Hydrocarbons (as C_2H_4)	100	0.5
Carbon monoxide	100	2
Sulfur dioxide	1	0.05
Ozone	0.5	0.01
Ammonia	10	0.002

Note: Maximum air flow: 100 standard ft^3/min. Desired air temperature: 72 ± 3°F. Desired air humidity: 35–95% RH, controllable. Particles: 99.97% removal for 0.3 μm diameter.

Source: Phalen, R. F., *Inhalation Studies: Foundations and Techniques,* CRC Press, Boca Raton, FL, 1984, 83.

that cools the air to about 35°F, which adjusts the water-vapor content of the air to a maximum saturation value at 35°F. An activated charcoal filter further reduces the levels of hydrocarbons. A 100-W electrical heater warms the air, and two additional adjustable 1500-W immersion heaters vaporize distilled water and add steam to the air to increase the humidity to the desired level. Automatic temperature and humidity controllers, or manual adjustments, are used to control the heaters to maintain the desired relative humidity in the exit air stream. Purified, conditioned air flows through a 10-cm pipe having a manually operated butterfly valve and passes through a manifold pipe system to the top inlet of four exposure chambers. The valve is used to adjust the total output of the air-purifying/conditioning unit. Each chamber is preceded by a HEPA filter and a butterfly valve. When desired, pollutants are injected into the air just downstream of the HEPA filter. Air flows downward through the chambers and exits through a prefilter and HEPA filter, followed by another butterfly valve. Four blowers on the roof pull air through the four chambers and exhaust it to the atmosphere. The butterfly valves for each chamber are used to adjust the air flows and chamber pressures.

B. A High-Pressure System

The second system (Figure 3.2) was designed by another engineer, Robert Walters, in 1979, after several years of experience with the previous system. Design criteria were similar, but a system that operated at higher pressure was sought in order to achieve lower humidities, to further decrease levels of ambient pollutants in the chambers, and to solve other minor problems. Outdoor air is first filtered and pressurized to 100 lb/in.^{2} using a liquid-ring (water) compressor (Nash Model CD663C; Norwalk, CN). Use of a reciprocating-type air compressor would have allowed higher system pressures (up to 5000 lb/in.^{2}) but would have meant a higher maintenance burden, reduced reliability, oil-vapor contamination of the compressed air, and the possibility of CO formation due to compressor lubricant breakdown. Further, Doyle et al.[6] had shown the liquid-ring compressor to be practical for use in high-pressure air purification systems. Water-saturated air leaving the receiver is filtered to remove water droplets and other particles and is passed through a 31-cm diameter by 80-cm deep bed of permanganate-impregnated alumina pellets (Purafil®). The unit efficiently removes SO_2, NO_x, O_3, and some hydrocarbons from the pressurized airstream. Further purification and drying of the airstream is next accomplished in activated alumina-filled heatless dryer columns (Deltech Engineering Corp.; New Castle, DE) operated in a pressure-swing desorption regenerating mode. In this operation, high pressure air is dried by passage through one drying column, while a portion of this dried air is simultaneously passed through a second idling column at atmospheric pressure to desorb moisture and contaminants. This enables the idling column to act as the drying column in the next cycle. Drying of the pressurized airstream to less than 0.1 mg/*l* water content improves the adsorption performance of the downstream activated carbon bed and is essential to prevent water vapor poisoning of the following unit, a fixed Hopcalite® bed (Deltech Engineering Corp.; New Castle, DE) operating at ambient temperature to remove carbon

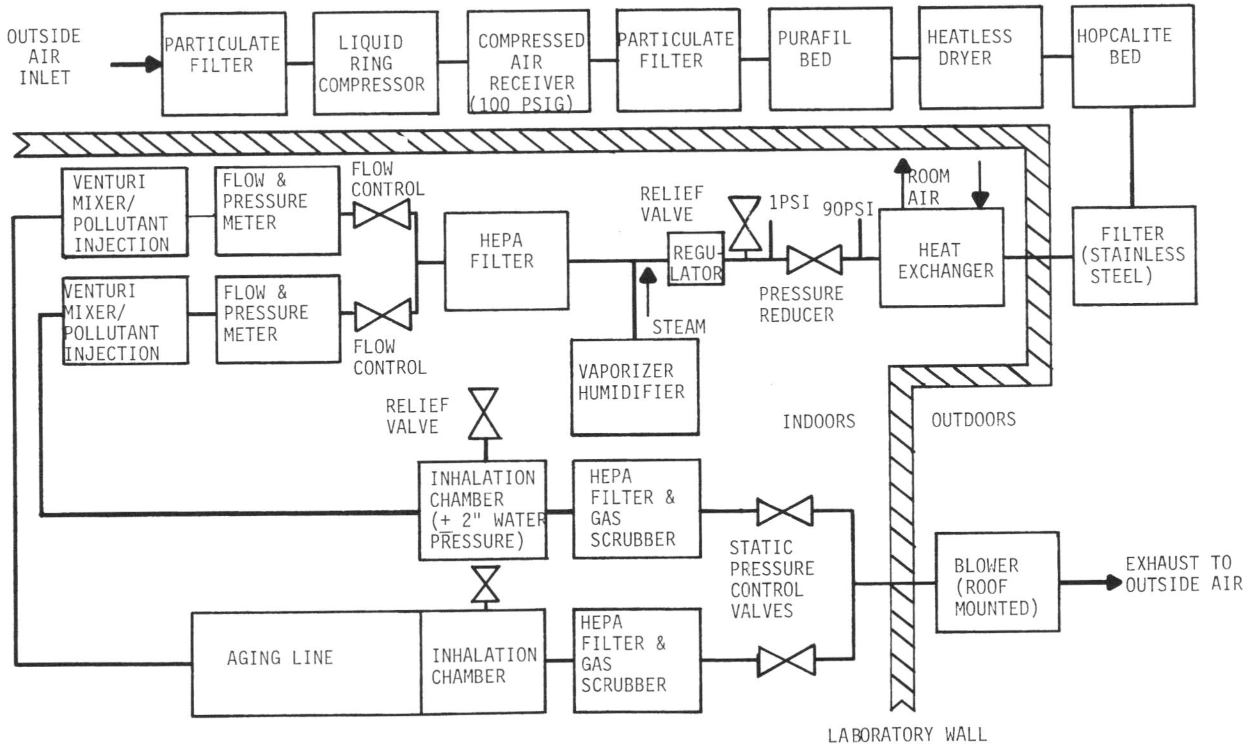

Figure 3.2
High-pressure air purification system used to supply two 1-m^3 volume chambers. (From Phalen, R. F., *Inhalation Studies: Foundations and Techniques*, CRC Press, Boca Raton, FL, 1984, 84.)

TABLE 3.6
Contaminant Removal by High-Pressure Air Purification System[a]

Compound	Ambient conc.	Purified air conc.	Measurement instrument
O_3	0–0.2 ppm	<0.01 ppm	Dasibi® Model 1003-AH Ozone Monitor
SO_2	0–1.0 ppm	<0.01 ppm	Teco® Model 43 SO_2 Analyzer
CO_2	0.04%	0.02%	Perkin-Elmer® #1100 Medical Mass Spectrometer
N_2	78.3%	78.3%	Perkin-Elmer® #1100 Medical Mass Spectrometer
O_2	21.0%	21.0%	Perkin-Elmer® #1100 Medical Mass Spectrometer
CO	0–30 ppm	<1.0 ppm[b]	Ecolyzer® Model 2000 CO Analyzer
NO	0.02 ppm	<0.01 ppm[b]	Beckman® Model 952 NO_x Analyzer
NO_2	0.08 ppm	0.04 ppm	Beckman® Model 952 NO_x Analyzer
CH_4	1.4 ppm	1.4 ppm	Beckman® Model 6800 Gas Chromatograph
C_2H_6	6.3 ppb	7.1 ppb	Beckman® Model 6800 Gas Chromatograph
C_2H_4	5.1 ppb	<2 ppb[b]	Beckman® Model 6800 Gas Chromatograph
C_2H_2	4.7 ppb	<2 ppb[b]	Beckman® Model 6800 Gas Chromatograph
C_3[c]	0.46 ppm	0.20 ppm	Beckman® Model 6800 Gas Chromatograph
Particulate	10^4–10^5 cc^{-1}	<100 cc^{-1}	Environment-One® Rich 100 Condensation Nuclei Counter

[a] Measured at system flow rates of 600 to 1200 std. liters min^{-1}.

[b] Lower detection limit for compound.

[c] Hydrocarbons with three or more carbon atoms.

Source: Phalen, R. F., *Inhalation Studies: Foundations and Techniques*, CRC Press, Boca Raton, FL, 1984, 85.

monoxide. This dryer/adsorption unit was designed originally for the purification of compressed breathing air (for scuba diving and other applications).

After passage through activated charcoal to remove heavier hydrocarbons, the airstream is passed through a filter to remove entrained dust and then is brought to indoor laboratory temperature using an air/air heat exchanger (similar to an automobile radiator in design). Delivery of the air to inhalation chambers follows air pressure reduction, humidification, and final high efficiency (HEPA) particle filtration. The pollution removal performance of the completed system for several common ambient pollutants was evaluated by monitoring ambient and processed air. The results are presented in Table 3.6. This performance is more than acceptable for conducting inhalation studies, even in an urban setting that has significant levels of air pollutants.

Useful air purification systems have been described by Doyle et al.,[6] Barsocchi and Knobel,[7] and Roehlich and Rodgers.[8] Doyle's system, a forerunner of that shown in Figure 3.2, was developed for purification and humidification of air for atmospheric chemistry research at the University of California, Riverside. Barsocchi and Knobel's design was intended for the filling of high-pressure cylinders of dry, zero (very low contaminant) air. The air purifier of Roehlich and Rodgers was developed for the purpose of removing automotive exhaust from the breathing air in working areas subject to heavy vehicular traffic. All of these systems are good basic designs upon which to draw for inhalation studies.

VII. Suppliers of Equipment

A. Gauges, Meters, Fittings, Tools, Controllers, etc.

McMaster-Carr Supply Co.
P.O. Box 54960
Los Angeles, CA 90054-0960

Fisher Scientific
711 Forbes Avenue
Pittsburgh, PA 15219-4785

VWR Scientific
P.O. Box 7900
San Francisco, CA 94120

Cole-Parmer Instrument Co.
625 East Bunker Court
Vernon Hills, IL 60061

B. Filters, Air Purifiers, Air Dryers, Gas Scrubbers

Deltech Engineering Corp.
Century Park
P.O. Box 667
New Castle, DE 19720

Airotech, Inc.
Boyle Center
120 Ninth Avenue
Homestead, PA 15120-1600

CSC
P.O. Box 3, Route 92 East
Bath, NC 27808

AirTech
2007 Raymer Street, Suite S
Fullerton, CA 92633

Purafil, Inc.
2654 Weaver Way
Doraville, GA 30340

Costar
1 Alewise
Cambridge, MA 02140

Filtra Corp.
104 Wagaraw Road
P.O. Box 298
Hawthorne, NJ 07506

Millipore Corporation
80 Ashby
Bedford, MA 01730

Mine Safety Appliances Company
Filter Products Division
P.O. Box 426
Pittsburgh, PA 15230

Gelman Sciences
Air Analysis Division
600 South Wagner Road
Ann Arbor, MI 48103

Deltech Engineering, Inc.
P.O. Box 667
New Castle, DE 19720

Motor Guard Corp.
580 Carnegie Steet
Manteca, CA 95337

VIII. Sample Data Obtained

The contaminant removal efficiencies for the Dennison- and Walters-design air purification systems were shown in Tables 3.5 and 3.6. Such data are excellent for air purification system performance.

References

1. Phalen, R. F., *Inhalation Studies: Foundations and Techniques*, CRC Press, Boca Raton, FL, 1984, chap. 3.
2. Rollins, J.P., Ed., *Compressed Air and Gas Handbook,* Fourth ed., Compressed Air and Gas Institute, New York, 1973.
3. Nelson, G. O., *Controlled Test Atmospheres*, Ann Arbor Science, Ann Arbor, MI, 1972.
4. Dorman, R. G., Filtration, in *Aerosol Science*, Davies, C. N., Ed., Academic Press, New York, 1966, chap. 8.
5. Burchsted, C. A., Kahn, J. E., and Fuller, A. B., *Nuclear Air Cleaning Handbook*, Energy Research and Development Administration Report, ERDA-76-21, available from National Technical Information Service, Springfield, VA, 1970.
6. Doyle, G. J., Bekowies, P. J., Winer, A. M., and Pitts, J. N., Jr., Charcoal-adsorption air purification system for chamber studies investigating atmospheric photochemistry, *Environ. Sci. Technol.*, 11, 45, 1977.
7. Barsocchi, A. T. and Knobel, R., Zero air: selection of a high-volume preparation method, *Am. Lab.*, 12, 81, Feb., 1980.
8. Roehlich, F., Jr. and Rodgers, S. J., Development of an air purifier for personnel cubicles, *Am. Ind. Hyg. Assoc. J.*, 37, 586, 1976.

Chapter 4

Exposure Atmosphere Generation and Characterization

Michael T. Kleinman and
Robert F. Phalen

Contents

0-8493-3343-1/97/$0.00+$.50

I. Introduction

The requirements of a modern inhalation study include having rigorous control of the exposure atmosphere (known concentrations of well defined components) and ensuring the absence of significantly confounding contaminants. To achieve these requirements, purified air with a controlled temperature and humidity is intentionally modified by added substances. Two basic considerations are (1) how to generate and control the exposure atmosphere, and (2) how to adequately monitor the exposure. This chapter discusses the selection and use of particle and gas generating systems, some of the problems which arise, and representive techniques for atmosphere generation and characterization.

A typical exposure system includes an air purification system, pollutant generators, the exposure unit, and the sampling systems. Peripheral systems include air flow controllers, air humidifiers/dryers, scrubbers, or filters to remove contaminants from air exiting the exposure system and a disposal system for water used to wash the exposure apparatus.

The generation of even simple atmospheres can pose significant problems above and beyond the selection and installation of the appropriate major equipment. For example, problems arise when exposure chamber or delivery line materials are not compatible with the atmosphere being studied. For example, the use of rubber (and some types of plastic) delivery lines leading from an ozone (O_3) generator can result in large losses of O_3. To overcome these losses, the generator must produce quantities of O_3 far in excess of those needed for the experiment. While this in itself is not a major problem, it could result in several complications, such as the co-generation of other reactive oxygen species along with O_3 and the contamination of the exposure atmosphere with volatile degradation products. Another problem arises with the use of nonconductive materials in the construction of delivery lines and chambers used for aerosol studies. The buildup of static charges on the walls of the tubes or the chamber can produce losses of particles to these surfaces and result in unstable particle concentrations. These problems are solved by the use of inert, electrically conductive materials.

II. Scope and Limitations

Since the number of materials studied in inhalation research is quite large, selected typical examples will be discussed. This chapter will focus on single component atmospheres, and complex aerosol/gas mixtures will be briefly addressed. There are too many methods for collecting, analyzing, and characterizing exposure atmospheres to permit an encyclopedic discussion. The benefits and limitations of several techniques, however, will be discussed.

III. Atmosphere Generation Techniques

A. Particles

Generation of atmospheres containing particles of known size and concentration can be especially challenging. Particle diameters of aerosols (particles suspended in gases) can range from 0.001 to over 100 μm. Monodisperse aerosols are often used in studies of respiratory tract deposition and clearance, for examining the influence of particle size on biological processes, and for directing delivery of materials to selected regions of the respiratory tract. Monodisperse aerosols contain particles of nearly uniform size and are generally defined as aerosols that have size distributions with a known median diameter and a geometric standard deviation of less than 1.2 (aerosol sizes are usually log-normally distributed). Selected methods for generating monodisperse aerosols include condensation aerosol generators (Sinclair-LaMer type generators), spinning disk atomizers, dispersion of liquid suspensions of monodisperse particles, and particle-size segregation.

Polydisperse aerosols (those which contain a broad distribution of particle sizes) are easier to generate and can be used when the characteristics of monodispersity are not necessary. Polydisperse aerosols are usually generated by nebulization (of solutions of soluble materials or suspensions of particles), suspension of dry powders or dusts by air jets, or by mechanical grinding or abrasion.

Generation techniques for monodisperse and polydisperse aerosols are discussed in several references.[1-6] Some of the more widely used methods will be presented here.

1. Monodisperse aerosols

Particles in a narrow size range can be obtained by the controlled condensation of organic vapors onto nuclei particles. Such devices were reported by Sinclair and LaMer,[7] and variations and improvements on this methodology have been summarized by Prodi,[8] Mercer,[1] and Hinds.[4] Generators of this type are commercially available. In the basic approach, a low vapor pressure liquid is heated in a boiler at a constant temperature. Nuclei particles, generated by any of several methods, are mixed with the condensable heated vapor. The aerosol is passed into a vertical chamber where it is slowly cooled, permitting the vapor to condense on the particles in a controlled manner. If the vapor is uniformly mixed with the nuclei and condensation is slow and diffusion-controlled, then monodisperse particles will be produced. By varying the flow rate of the nuclei particles through the vapor and the vapor temperature, particles having count-median diameters ranging from about 0.2 to 2.0 μm can be produced. The geometric standard deviation of the particles generated by this type of device is generally less than 1.1.

In a paper published in 1949, Walton and Prewett[9] presented an analysis of the conditions for production of monodisperse particles that are thrown off of the edge of a rapidly spinning surface. Since that time, a variety of spinning-disk and spinning-top liquid aerosol generators has been described. These generators operate at rotational speeds ranging from a few hundred up to about a quarter million revolutions per minute, and produce primary monodisperse droplets with diameters in the range of 6 μm to 3 mm. (They also produce small secondary satellite particles that form as the main droplets separate from the edge of the spinning surface; these particles must be removed.) Subsequent drying of primary particles can be used to produce monodisperse particles as small as 1 μm in diameter. Lippmann and Albert[10] designed a spinning-disk generator for producing radioactive insoluble aerosols for human and laboratory animal deposition and clearance studies. In this generator, satellite particles are separated from the main particles and collected on a filter. Precise control of the liquid feed to the spinning disk is necessary to achieve monodispersity. Optimal conditions for using the spinning-top generator to produce monodisperse aerosols are discussed by Mori et al.[11]

A relatively simple method for producing monodisperse aerosols in a wide variety of sizes involves the aerosolization and drying of aqueous suspensions of commercially-available latex spheres (available from Duke Scientific Corp.; Palo Alto, CA). These spheres are supplied in diameters ranging from less than 0.1 μm up to over 100 μm. The particles can be labeled with fluorescent dyes, radioactive elements, or magnetic materials. Biologically active substrates also can be attached to these particles. Suspensions of these particles can be aerosolized using a variety of generators; compressed-air nebulizers are frequently used. Although the commercial particles are quite uniform, several considerations must be applied in their use. First, the nominal diameters, as supplied by the manufacturer, are based on microscopic measurements, but the aerodynamic behavior may depend on conditions in a specific investigation; therefore, aerodynamic sizes may need to be measured. When liquid suspensions are aerosolized, some droplets will contain more than one solid particle, and some will contain no particles but will contain residual contaminants (salts, surfactants, etc.). Thus, after drying, in addition to individual monodisperse particles, some agglomerates (two or more particles stuck together) and some small residue particles will be components of the aerosol. As the aerosol is made more concentrated, the agglomerate fraction increases, and as it is made more dilute, the residue fraction is greater. Raabe[12] has given dilution factors for suspensions that allow one to keep the singlet/agglomerate ratio above a given value. As an example, using a nebulizer with a volume median droplet diameter of 5 μm and a geometric standard deviation of 1.6 to aerosolize a suspension of 1.0-μm diameter latex spheres, the percent solids (by volume) in the nebulized suspension should be below about 0.03 (i.e., 1 part in 3000) in order to keep the ratio of singlets to agglomerates above 20:1. In this case, the ratio of small stabilizer residue particles to single latex particles will be about 50:1. Monodisperse powders can also be resuspended in air using suitable fluidized bed-type generators.[13]

Monodisperse droplets can be made in relatively small quantities by the vibrational breakup of a liquid stream.[14-16] In some devices, a liquid stream exiting from a small orifice (10 to 15 μm in diameter) is broken into droplets by a vibrating piezoelectric crystal. Vibration of the orifice at frequencies ranging from 50,000 to 500,000 Hz (cycles per second) leads to droplets of sizes from 18 to 40 μm that vary in volume by less than 1%. A vibrating orifice, monodisperse aerosol generator is commercially available (Model 3050, Thermo Systems Inc.; St. Paul, MN), after the design by Berglund and Liu.[17] Modifications which improve the reliability of this type of generator have been reported by Mitchell et al.[18]

2. Polydisperse aerosols

A great variety of techniques has been used for controlled generation of polydisperse aerosols. Even a simple listing of these techniques would be cumbersome, but certain methods are especially useful because they produce aerosols in the size ranges similar to those inhaled by people in various situations. The source material for the aerosol can be a liquid (solution or suspension) or a solid. Liquids can be aerosolized using air blast or ultrasonic nebulizers. Examples of generators for solid substrates include the Wright dust feeder for dispersion of dry compressed powders, the air-blast nebulizer for aerosolization of liquids, the fluidized-bed generator for dispersion of loose powders, the exploding-wire generator for vaporizing solid metals, and the combustion generator for generating fuel-derived particles. Several references treat methods for generating polydisperse particles.[1-6] A few methods will be presented here.

Droplet aerosols are used in several ways in inhalation studies: (1) the droplet itself may be passed into an exposure system for direct inhalation; (2) the droplets may carry a suspended solid that is inhaled after evaporation of the droplet; (3) a solution may be aerosolized, leaving a residue particle upon drying; and (4) the droplets may be used as input to another stage of an aerosol generator (for example as input to the heater of a condensation aerosol generator). Air-blast and ultrasonic nebulizers produce polydisperse droplets suitable for many scientific purposes (Table 4.1). Each generator has its unique characteristics and any one may be preferable over the other for a given application. If a study requires a high number of droplets per liter of air and

TABLE 4.1
Two Polydisperse Droplet Aerosol Generators

Type	Operating principle
Air-blast nebulizer	High-velocity air directed over a liquid feed tube produces breakup of entrained liquid filament
Ultrasonic nebulizer	Vibration of a piezoelectric crystal forms a fountain of liquid that emits droplets from its tip

Source: Phalen, R. F., *Inhalation Studies: Foundations and Techniques*, CRC Press, Boca Raton, FL, 1984, 89.

large droplets can be tolerated, an ultrasonic nebulizer may be used. Compressed air nebulizers are useful for providing smaller droplets at somewhat lower output rates.

Compressed air nebulizers are commonly selected for inhalation studies due to their relative low cost, stable operation, and droplet size characteristics. Table 4.2 presents the operating characteristics of some commonly used compressed air nebulizers. In this type of generator, an air jet shatters a liquid stream which is drawn from a reservoir into the path of the air jet by the reduced pressure within the jet or as a result of positive pressure in a reservoir. The shattered liquid forms large and small droplets, and the larger droplets impact on a nearby surface (an adjacent wall, a ball, a baffle, etc.). Smaller droplets follow air streams around the impaction surfaces and exit from the nebulizer. The air that entrains the particles will be saturated with the vapor of the fluid in the reservoir, so the concentration of the dissolved or suspended material may steadily increase with time. Three methods are used to control this effect: (1) the reservoir may be cooled to reduce evaporation; (2) the reservoir may be enlarged or continuously renewed; or (3) the feed air can be pre-saturated with liquid vapor. Although the majority of the droplets impact within the nebulizer, usually about 10^6 to 10^7 particles per cubic centimeter of air exit as useful aerosol. These droplets typically have a distribution of sizes that is describable by a log-normal distribution function with a geometric standard deviation between about 1.3 and 2.2. Freshly nebulized droplets usually are highly electrically charged and are often passed through a bipolar ion field for discharging.[19]

When a solution is nebulized and the droplets dried to a residue aerosol, the resultant dry particles will be smaller than the original droplets. The theoretical relationship between the mass median droplet diameter (D_d) and the mass median residue particle diameter (D_p) for spherical residue particles is

$$(D_p)^3 = (D_d)^3 \, (C) \, (\rho_d/\rho_p) \tag{4.1}$$

where ρ denotes density (subscripts relate to the droplet, d, and to the residual particles, p) and C is the mass fraction of solute. The geometric standard deviation of the residue aerosol will be approximately that of the droplet aerosol.

TABLE 4.2
Operating Characteristics of Compressed Air Nebulizers

Nebulizer	Applied pressure (lb/in.^{2})	Air flow (l/min)	Aerosol output conc. (g/m^3)	Vol. median drop diameter[a] (μm)
3-Jet Collison	15	6.1	8.7	—
(available from BGI, Inc.)	30	9.4	9.0	—
	50	13.6	10.4	—
De Vilbiss® No. 40	15	12.4	15.5	4.2
with closed vent	20	16.0	14.0	3.2
(available from De Vilbiss)	30	20.9	12.1	2.8
Lovelace	20	1.34	(34)[b]	6.9
(available from In-Tox)	30	1.81	(22)[b]	4.7
	40	2.28	(15)[b]	3.1
	50	2.64	(19)[b]	2.6
(Nebulizer chilled to 0°C)	20	1.34	55	—

[a] G.S.D. (geometric standard deviation) for these methods ranges from 1.6 to 2.3.

[b] Calculated.

Source: Phalen, R. F., *Inhalation Studies: Foundations and Techniques,* CRC Press, Boca Raton, FL, 1984, 90.

TABLE 4.3
Selected Dry Dust Aerosol Generators

Generator	Operation principle
Wright dust feed	Powder which has been compressed into a block is scraped by a rotating blade and entrained in a dry air stream.
Timbrell-type fibrous dust generator	Similar to Wright dust feed, but blades rotate at high speed over compacted asbestos or fibrous glass plug.
Fluidized bed	Upward flowing air fluidizes (suspends) a bed made of large (180 μm) "beads" into which dust to be aerosolized is mixed. The air carries particles upward into a vertical elutriator. New powder is continuously fed into the fluidizing section.
Turntable dust feed	Powder flows from a hopper (agitated) onto a turntable, which rotates toward an aerosolizing air jet.

Source: Phalen, R. F., *Inhalation Studies: Foundations and Techniques,* CRC Press, Boca Raton, FL, 1984, 91.

Table 4.3 lists some useful dry dust aerosol generators. The dust loaded into the generator can have a variety of forms, including fibers, spheres, plates, etc. The dispersed phase will have a size distribution that may differ from that of the bulk dust material, due to breakup or agglomeration of primary particles.

The fluidized-bed elutriator is often used in inhalation studies because of its stability over long periods of generation and its favorable elimination of larger agglomerates and large primary particles by elutriation (settling downward). Modern fluidized-bed elutriators usually incorporate two important features: continuous feeding of material to the generator and electrical discharging of the aerosol within the elutriation column. Continuous feeding of material into the generator is often necessary because material in the generator initially loses the fine particle fraction, leaving the larger particles in the generator. Thus, unless material is continuously supplied to the generator, the emitted size distribution will change with time. Fluidized bed generators impart electrical charges to the aerosol they emit, and thus passage through an bipolar ion field is recommended.[20] The relationship between fluidized bed operating conditions and the aerosol produced has been reviewed by Lua.[21] Key factors are the fluid bed height and the airflow through the bed after the minimum fluidizing velocity is exceeded.

The dust feeder described originally by Wright[22] has been used to generate aerosols from a wide variety of dry materials. The material is first formed into a plug, which is then scraped by a blade and aerosolized by entrainment in a flowing air stream. An important limitation of this generator is that the material must pack into a stable plug with sufficient integrity to avoid fragmentation during the scraping process.

3. General considerations

Selection of appropriate aerosol generators for inhalation studies is made on the basis of several criteria. Some of the more important considerations in aerosol generator selection include the following:

1. Are the aerosol characteristics appropriate? These characteristics include chemical composition and physical properties such as shape, size, surface area, dispersity, and number of particles per unit volume of air.
2. Is the generator and its aerosol stable over the time period of intended use?
3. Does the generator waste input materials? Some materials that are aerosolized are expensive or rare, and economy of operation can be important.
4. Is the generator hazardous to laboratory personnel or experimental subjects?
5. Does the generator introduce unwanted foreign materials into the aerosol?
6. Does the generator require a large amount of maintenance?
7. How sensitive are the characteristics of the output aerosol to perturbations in inputs or environmental parameters such as voltage, input air pressure, barometric pressure, temperature, and relative humidity?
8. Has the aerosol generator been described and/or evaluated in publications in the available literature?

Persons not experienced in aerosol generation must be very careful when selecting generators for inhalation studies. When in doubt, one is wise to contact aerosol scientists to discuss one's requirements.

B. Gases

The generation of gases usually involves the controlled introduction of a concentrated gas into a stream of purified air. The mixture is then introduced into a chamber or other dynamic exposure system. Alternatively, the gas is injected into a static system, which may be useful for certain purposes, such as measuring respiratory system uptake. The most widely used methods of generating gases include metering from pressurized cylinders, diffusion of the gas through a permeation tube source, vaporization of volatile fluids, and creation of the gas by controlled reactions.

In some respects, the generation of gases is less difficult than the generation of particles; stabilization of particle size is not a consideration, and usually it is easy to achieve uniform mixing of a gas in an airstream. Also, the devices used to control the injection of a gas into dilution air are comparatively reliable and simple; however, the gas must be of high purity. It is essential that the materials used in the construction of the generator, metering device, and delivery lines not introduce contaminants (including particles), cause chemical degradation of the gas, or otherwise consume the gas. It is frequently necessary to pre-condition the generator and delivery and exposure systems so that stable concentrations are achieved for the exposures.

1. Pressurized cylinders

Cylinders containing pure gases (or mixtures of gases) at pressures up to a few thousand pounds per square inch (psi) are available commercially or can be prepared in the laboratory. Cylinders are prepared by first evacuating them to a sufficiently low pressure to remove contaminants, introducing a known amount of the desired gas, and then filling under pressure with a carrier gas (usually purified air or nitrogen). Mixing a filled cylinder can be achieved by rotating the cylinder or, alternatively, by gently heating and cooling an end of the cylinder. While some gases may be stable for long periods when stored, others require cylinders made with nonreactive materials (e.g., stainless steel or aluminum) or the use of inert carrier gases (e.g., nitrogen or helium) to extend shelf life.

Commercially available gases of extremely high purity are available; however, some gases will spontaneously decompose over time. For example, nitrogen dioxide (NO_2) under pressure at high concentrations (>1%) will invariably contain a small amount of nitric oxide (NO), a dissociation product. In addition, carrier gases can contain small amounts of contaminants such as hydrocarbons, carbon dioxide, and water vapor. In most inhalation studies, the pressurized gases are diluted many-fold, and the resulting amount of contamination at the breathing zone is negligible. In those cases where even a low level of contamination is a problem, scrubbers can be used to reduce contaminant concentrations. Gases for administration to humans and that conform to U.S. Pharmacopeia (USP) standards can be obtained commercially.

The cylinders from commercial suppliers generally have a valve with a threaded outlet which conforms to a substance-specific standard thread which fits only specific pressure regulators. This practice helps to minimize the possibility of mixing incompatible gases (e.g., hydrogen with oxygen), or of contaminating a system with gas residuals in a pressure regulator previously used with another gas.

Pressure regulators are used to reduce the pressure of the compressed gas to a safe working level. Two-stage regulators are recommended for most purposes. The first stage reduces the pressure from that in the tank to a pre-set intermediate level. Because the pressure fed to the second stage is fixed by the first stage, the delivery pressure is not affected by changes in the cylinder pressure, thus providing precise control of gas delivery during the course of a study. Regulators for delivery of high purity gases, or corrosive gases, are constructed of high-quality stainless steel (e.g., 316-grade stainless), with stainless steel or fluorocarbon-lined diaphragms and fluorocarbon or stainless steel seats and seals. Prices for regulators can range from about $200 for general models to about $1000 for delivery of ultra-high-purity, or corrosive, gases.

To control exposure concentrations precisely it is necessary to carefully meter the gas into the airstream by using a flow controller. Flow controllers can be a fixed-flow type (critical orifices, capillary tubes, or porous plugs) or variable-flow type (needle-valve-equipped rotameters, orifice meters, heated wire anemometers, or electronically controlled mass-flow meters). Flow controllers should be calibrated using devices such as a bubble meter, a spirometer (primary flow calibrators), or a previously calibrated wet-test or dry-gas meter (a "transfer" standard).

Compressed gases are often purchased for inhalation studies at very high concentrations. Small leaks from the delivery system into the laboratory can potentially result in relatively high exposures of personnel. All connections should be checked carefully for leaks, all cylinders should be properly secured, and, if possible, periodic area monitoring should be performed. Although it is a rare occurrence, regulator diaphragms can crack, releasing high pressure gas into unsealed portions of the regulator and then into the room. Therefore, two-stage regulators for toxic gases are equipped with a pre-set safety valve that protects the second stage from overloading and can be connected to a vent line to safely exhaust released gases.

2. Permeation tubes

Permeation devices are polymeric tubes that are permeable to vapors of the liquids which are sealed within them. Hundreds of organic and inorganic chemicals are commercially available in permeation tubes. The permeation rate of the gas out of the tube is a function of the vapor diffusion coefficient, vapor pressure, temperature, atmospheric gas pressure, solubility coefficient of the vapor in the tube polymer, surface area of the tube, and the thickness of the tube wall. The vapor permeation rate is determined by weighing the tube at timed intervals. In use, the permeation tube is inserted into a special chamber which is flushed with purified air at a known, constant rate. The chamber is maintained at a fixed temperature by immersion in a constant temperature bath or by use of a thermostated heater. The vapor-laden air which exits the permeation tube system can be diluted with a known flow of purified air to produce a stable delivery rate. Permeation tubes can be purchased or constructed, in various sizes and with various permeation rates. For heat-stable materials, the temperature of the permeation tube chamber can be increased to increase the permeation rate. Multiple permeation tubes can be used in parallel to increase output concentrations, if necessary.

Some precautions are necessary to ensure good results when using permeation tubes. Careful handling of the tube is necessary because damage to the tube or the presence of finger oils or other contamination can alter the permeation rate. The tube should never be handled without appropriate gloves and should only be handled by the sealed ends to avoid damaging the permeable surface of the tube. A temperature-controlled chamber and an accurate balance are needed to maintain and to check the permeation rate. Note that even at room temperature, permeation tubes continue to emit vapors; hence, tubes should be stored between uses at a low temperature in a sealed container. The container should only be opened in a well ventilated area, preferably an exhaust hood, to avoid exposure of personnel.

3. *In situ* generation by physical or chemical reactions

Several physical and chemical processes lead to the production of gases. Such processes can be exploited when the gas to be generated is not commercially available or if the gas is unstable. Physical process that lead to the generation of gases include thermal transformations, photochemical reactions, and electrolysis from molten salts.

Chemical reactions that produce gases can be used to produce compounds that otherwise would be unavailable. The rates of production depend upon a number of factors that must be rigorously controlled in order to maintain a stable output with respect to purity and concentration. Reaction rates depend upon thermodynamic factors such as temperature, pressure, and humidity, as well as kinetic factors such as the concentrations of the reactants, the rate of reactant mixing, and the rate of removal of products. Impurities in the reactants or in the reaction vessel may lead to contamination of the product or modification of the reaction rates and pathways. In some situations the evolved product, which might be obtained in a liquid or vapor form, can be collected, purified, and metered into the exposure system using metering pumps or motor driven syringes.

As a specific example of a gas generated by physical processes, consider O_3. To produce O_3, diatomic oxygen (O_2) should be used as a starting material, because the use of air could result in production of unwanted contaminants (such as nitrogen oxides). A laboratory O_3 generation system might consist of a cylinder of compressed pure O_2, a two-stage pressure regulator, a needle valve-equipped rotameter to meter O_2 flow to the reactor, and a voltage-controlled electrical discharge or ultraviolet reactor. The $O_2 + O_3$ mixture should exit the reactor through preconditioned inert tubing prior to being mixed with purified air and injected into the exposure system. Concentrations of O_3 in the exposure system should be monitored with a continuous monitor and adjusted as needed by controlling the flow rates or the voltage applied to the reactor. Depending upon the metering rate of O_2 into the generator and the exposure system air throughput rate, the chamber O_2 concentration could be higher than ambient, which might be important if metabolic measurements are being made as part of the study. Nitrogen gas can be metered into the chamber to adjust the concentration of O_2 to normal ambient levels.

4. Evaporation or sublimation

Controlled generation of vapor by evaporation or sublimation can be useful for generating some gases (such as formaldehyde or nitric acid vapor). The rate of gas evolution is temperature dependent, so precise control of the generator temperature is required. Immersion of the reactor vessel containing the liquid or solid into a thermostatically controlled bath will often suffice. Carrier gas for the generator can be heated to the required temperature by passage through a metal coil immersed in the same bath. Many vapors at high concentration are reactive, explosive, or flammable. In such cases, an inert carrier gas, such as nitrogen, should be used. If it is necessary to use air, the vapor should be rapidly diluted to a concentration below the lower explosive limit (LEL); for safety, a maximum concentration of 10% of the LEL should be considered.

IV. Characterization Techniques

A. Introduction

Careful characterization of the exposure is a necessary part of experimental inhalation toxicology studies and thus deserves both appropriate planning and attention to detail. The usual objective in an aerosol inhalation study is to establish a quantitative relationship between an observed biological response and the measurable physical and chemical properties of the test aerosol. The relevant physical and chemical properties that produce the expected biological effects must be known. Certainly the composition, number and mass concentrations, median size, and size distributions come immediately to mind. Other particle properties, including the surface area, state of electrical charge, surface character, hygroscopicity, particle shape characteristics, dissolution rates in lung fluids, and amounts and types of adsorbed materials often will be relevant to the study. Selection of the exact parameters for measurement is not trivial. Consider the size parameter. Which sizes are important — geometric, aerodynamic, or both? Should size distributions be based on particle number, surface area, volume, or mass? For a size distribution, is the mode, mean, or median most relevant?

Although gases are usually easier to characterize than particles, some issues must be considered. How short should the sampling intervals and instrument response times be? Should concentration data be summarized by the mean and standard deviation or by the median and geometric standard deviation? Should the peak value be recorded? The answers depend upon the study objectives.

Assuming that one has identified those relevant physical and chemical properties and has selected an appropriate battery of equipment, one is faced with additional questions. Where should one sample? How many samples must one acquire? A good inhalation study may involve expending as much effort in atmospheric characterization as is spent in measuring the biological response. In the following sections, some counsel will be offered on atmospheric characterization, but no simple or certain rules can be given that can be applied in all cases.

B. The Breathing Zone

Usually, one wishes to characterize the material in an inhalation study as it exists just before it is inhaled by a subject. Because it is often impractical to place a sampler inlet beside the subject's nose, the concept of the breathing zone will be introduced. The breathing zone is that volume or space from which the subject breathes. It extends from above the lowest to below the highest elevation commonly reached by the subject's nose during the exposure, and laterally it includes the full range of travel of the head. For a caged, unrestrained rat this may be the entire volume of the cage. For the masked animal, it will be the small volume in front of the nostrils. The major point to be made is that the breathing zone is not above or below the cage, it is not 2 ft into a pipe leading to a mask, nor is it at the outlet of the aerosol generator.

Implicit in this discussion of the breathing zone is the assumption that this zone has a uniform composition. If it does not, one should consider either additional confinement of the subject, better mixing, or sampling at sufficient sites so that an average composition can be described. It should be noted that subjects may preferentially breathe from areas in the breathing zone that contain less toxicant.

C. What Should Be Measured?

All of the particle properties that significantly contribute to biologic effects in the study should be measured. Table 4.4 lists 21 candidate properties that should be considered for measurement. Properties 1 through 5 nearly always must be considered. Property 6 (electrical charge) is often conveniently controlled by use of bipolar ion sources for reducing the state of electrical charge. In specific cases, the other listed properties may be important. For example, sodium chloride particles, upon entering the elevated relative humidity of the respiratory airways, become liquid and exhibit hygroscopic growth (the 7th property), which can alter the amount and distribution of material deposited. A well known example of the importance of property 9 (particle shape) is asbestos. Such particles, by virtue of their extreme length, can be retained in the deep lung for long periods. Practical considerations will limit the number of particle properties that can be measured in any given investigation, so good judgment on the part of the investigators is essential to the success of the sampling strategy.

The problem of characterizing gaseous atmospheres is relatively straightforward in comparison to particles. In general, one must know either the concentration in terms of volume (e.g., ppm), or mass per unit volume of air (e.g., mg m^{-3}), and the total atmospheric pressure, or the partial pressure (torr, mmHg, etc.). As discussed for aerosol particles, gas concentrations must be measured in the breathing zone. Although gases tend to distribute more uniformly than particles in an air volume, local inhomogeneities in concentration may exist. Causes for this include the presence of sinks, such as reactive materials in the exposure system and leaks in the containment system. A factor which is often overlooked in characterizing a gaseous atmosphere is the determination of contaminant particles that may be present. In general, despite the precautions taken, some particles will be present which may influence the biologic effects of the gas that is being studied. For example, the results of a study that indicated a synergistic toxic effect between ozone and sulfur dioxide are now thought to be due to contaminant particles formed by gas-phase reactions between the added gases and the unpurified air used for the primary chamber throughput air.

Transient concentration spikes (or peaks) may be important in interpreting studies of gas toxicity. For example, peak ambient concentrations may be very important in observed health effects. In view of this, it is wise to record concentration maxima in inhalation studies. This type of monitoring implies the use of instruments which continuously sample, have a short cycle time, and can provide frequent concentration data (often by using an automatic data recorder). A sufficient number of samples should be taken so that the standard deviation of the gas concentration during the exposure can be calculated and reported along with the experimental results.

The toxicity of essentially all inhaled materials is influenced by the prevailing environmental conditions during exposure (Table 4.5). Any environmental parameter that produces a change in the subjects' breathing or ventilation rates will influence the dose received. Environmental exposure conditions which stress the subject can alter the susceptibility to injury, even with a given dose. Table 4.6 lists some of the most important environmental contaminants found in exposure chambers. Such contaminants should be controlled

TABLE 4.4
Aerosol Particle Properties that May Relate to Their Biological Effects

	Property	Comment
1.	Chemical composition	Will influence response in many ways
2.	Mass concentration (mass of particles per unit air volume)	Will influence dose deposited and, hence, biologic response
3.	Aerodynamic size	Will influence inhalability and deposition pattern
4.	Size distribution	Will influence dose and dose distribution
5.	Geometrical size	Will influence deposition pattern, clearance rate, dissolution rate, and sensory irritancy
6.	Electrical charge	Will influence coagulation rates and deposition on chamber walls, animal fur, nares, and nasal hairs
7.	Hygroscopicity	Will determine growth rates in respiratory tract and, hence, deposition pattern
8.	Surface area	Will determine amount of adsorbed gases carried on particles
		Will influence rate of dissolution in lung fluids; may influence interaction with macrophages and other lung cells
9.	Particle shape	Will influence deposition via interception mechanism and may influence phagocytic efficiencies
10.	Dissolution rate	Will determine persistence in lung fluids and tissues
11.	Water solubility	Will influence particle stability in respiratory tract
12.	Deliquesence	Will influence particle size in respiratory tract and, hence, deposition pattern
13.	Irritancy	May influence breathing patterns and thus modify dose and dose distribution
14.	Specific gravity	Will influence deposition pattern and may influence clearance rates
15.	Number concentration (no. of particles/volume of air)	May influence deposition pattern via "cloud effect" and will influence irritancy
16.	Antigenicity	May influence irritancy, induce tissue responses, and secondarily influence clearance rates
17.	Odor	May induce avoidance behavior
18.	Taste	May induce avoidance behavior
19.	Radioactivity	Can influence electrical charge and dissolution rate characteristics, as well as toxicity to lung cells
20.	Surface characteristics	May influence the interactions with macrophages and other lung cells
21.	Temperature	Particle temperature will be similar to that of surrounding gas and will influence deposition pattern

Note: Many of these properties are interdependent; for example, for spherical particles, the aerodynamic diameter is a function of geometric diameter and density.

Source: Phalen, R. F., *Inhalation Studies: Foundations and Techniques*, CRC Press, Boca Raton, FL, 1984, 104.

TABLE 4.5
Environmental Parameters Which Influence Subject Responses in Inhalation Studies

Parameter	Comments
Temperature	Will influence physical activity, ventilation rates, and several other biologic characteristics
Relative humidity	Influences thermal regulation and hence ventilation; extreme values can produce abnormalities in respiratory tract mucus
Atmospheric pressure	Influences ventilation as well as cardiac and hematopoietic functions; is directly related to driving pressure for uptake of gases by tissues
Illumination	Influences state of activity and can cause stress due to altered diurnal patterns
Air contaminants	Can act as co-toxins or lead to adapted or sensitized states; may influence ventilation and other physiologic functions
Noise	A known co-stress for mammals; often present in exposure chambers due to human activity, equipment vibrations, or flow of air through small orifices or over sharp edges
Vibration	A known co-stress that may be produced by motors or other moving apparatus

Source: Phalen, R. F., *Inhalation Studies: Foundations and Techniques*, CRC Press, Boca Raton, FL, 1984, 106.

TABLE 4.6
Common Contaminants Found in Exposure Chambers During Inhalation Exposures

Contaminant	Comments
Ammonia	Produced by bacterial action on urine; levels can rapidly increase over the duration of an exposure
Carbon dioxide	Crowding of animals or low airflows can cause a buildup of this exhaled metabolic byproduct
Water vapor	Crowding of animals or low air flows can lead to rapid buildup
Oil mist	Present in unfiltered output air from oil-lubricated air pumps
Animal dander and fur	May be present in excessive amounts due to poor housekeeping, overcrowding, or unhealthy animals
Organic vapors	May be residual from cleaning agents or previous exposures
Viable aerosols	May be present due to inadequate chamber cleaning

Source: Phalen, R. F., *Inhalation Studies: Foundations and Techniques*, CRC Press, Boca Raton, FL, 1984, 106.

and/or measured. The importance of proper control of environmental parameters cannot be overemphasized. Unless the investigator is aware of the exposure environment under the conditions of an actual exposure, he or she is not in control of the study.

D. Aerosol Characterization

The selection and use of an appropriate array of aerosol samplers and sizing instruments are not trivial tasks. Instrument limitations are not always obvious, and untrained or inexperienced persons can be at the mercy of the claims made in advertisements and manufacturer's literature. Aerosol characterizations represent some of the most difficult of all physical measurements, and even persons with advanced degrees in physics or engineering may not have had sufficient training to undertake them with confidence. It can be safely stated that no instrument, in the hands of a novice, will consistently provide reliable aerosol data.

Before discussing aerosol sizing instruments, it is necessary to consider devices that collect aerosol samples for subsequent analysis. Several sample collection devices are listed in Table 4.7. A basic sampling method, filtration, will be discussed briefly. An air filter is a device that allows the passage of air but traps particles. Collection efficiency varies with filter type, particle size, and the velocity of air passing through the filter. To collect particles, air is drawn through a filter using a pump or other method for producing a reduced pressure at the collector outlet, thus producing an air flow. Mercer[1] has extensively reviewed air filtration and has tabulated examples of filter collection efficiencies. Filters are usually of two types: pads of compacted fibers or membranes having many holes (pores) penetrating through them. Fiber filters designed for particle collection usually have fiber diameters in the micrometer size range. The fibers can be composed of a variety of materials including glass, viscose, cellulose acetate, asbestos, cotton, or other mineral and organic substances. Fibers can be coated with other materials, such as fluorocarbons, to modify chemical reactivity and hygroscopicity. Membrane filters can be made of cellulose or polycarbonate esters having pores of controlled diameter between about 0.01 and 10 μm. Membrane filters typically offer considerably more resistance to air flow than do fiber filters, but microscopic examination is easier for membrane filters, and they often have lower background concentrations of potential contaminants such trace metals or sulfates.

Filter samples taken for gravimetric analysis of airborne particle mass concentrations must be obtained with great care to avoid artifacts. Such artifacts include: changes in filter collection efficiency or throughput air flow rate due to clogging by the captured particles, leaks around the filter or through small tears, gain or loss of deposited mass due to chemical reactions, errors in determining filter mass due to loss of pieces of filter during handling, errors due to loss of sample in handling, and nonrepresentative sampling due to improper flowrate. From this list it is apparent that even this most basic method of obtaining a particle sample requires a knowledge of the physics of air flow and particle collection, the possible chemical reactions that can occur on the filter, and other relevant factors that may be specific to the application at hand. When care is taken, filter samples can provide accurate estimates of aerosol mass concentrations and the chemical species present.

TABLE 4.7
Aerosol Sample Collection Devices for Subsequent Analysis of Aerosol Shape, Size, or Composition

Sampler	Principle	Artifacts and limitations
Filters	Air passes through holes, channels, or around fibers, but particles are trapped by interception, impaction diffusion, sedimentation, or electrostatic interaction	Flow rate may change as filter loads, air flow may be nonisokinetic, filter media may be damaged, filter holders may leak, and reactions may occur on filter
Electrostatic precipitators	Particles are charged and attracted to a collection surface	Very tiny particles may lose charge
Thermal precipitators	Particles are driven onto a relatively cool collection surface by thermal gradient forces	Aerosol particles may be altered by high temperatures; condensation may alter particle size
Elutriators	Particles settle under the action of gravity onto a collection surface	Particles smaller than about 1 μm in diameter do not readily settle because of Brownian motion
Impingers	Aerosol is passed through a tube and impacted on a plate submerged in a liquid	Collection efficiency diminishes for particles smaller than about 1 μm in diameter; agglomerated or fragile particles may break up; soluble components may be aerosolized
Cyclones	Particle-laden air is given a rotational motion, leading to inertial collection of particles	Collection efficiency diminishes for particles smaller than about 5 μm in diameter; plastic cyclones develop charges and alter size distributions

Source: Phalen, R. F., *Inhalation Studies: Foundations and Techniques,* CRC Press, Boca Raton, FL, 1984, 108.

Aerosol measuring instruments are based on exploiting size-dependent properties of aerosols (Table 4.8). Examples of size-dependent aerosol properties include terminal settling velocities, light-scattering patterns, mobilities, and abilities to serve as centers (nuclei) for condensation of supersaturated vapors. It is a well known fact in physics that detection or measurement is always associated with some perturbation of the system under measurement. This perturbation may be negligible for a macroscopic body but may be appreciable for a tiny aerosol particle. Energy from an intense light beam, for example, can be absorbed and converted to heat, which can produce wild gyrations in the motion of tiny particles and may rapidly evaporate liquid droplets.

Two very important aerosol size distribution parameters, the mass median aerodynamic diameter and the geometric standard deviation, are usually reported. A calibrated cascade impactor with a backup filter is probably the most widely used instrument for obtaining these parameters. The use of cascade impactors and step-by-step instructions on data reduction have been treated extensively by Lodge and Chan.[23] Assuming one has properly taken an impactor sample and avoided the many possible artifacts listed in Table 4.8,

TABLE 4.8
Aerosol Measuring Instrumentation for Inhalation Studies

Instrument	Principle	Artifacts and limitations
Optical microscope	Image formation using visible light and glass optics	Lower limit of resolution = 0.5 μm can lead to grossly distorted size distributions
Electron microscopes	Projected image formation using electron beam and magnetic "optics"	Evaporation of sample due to vacuum and local heating; shallow depths of focus and small field of view can bias size distributions; coating of small particles with column contaminants can lead to overestimate of size
Scanning electron microscope	Image reconstruction using scanning electron beam scattering and electronic signal processing	Evaporation of sample due to vacuum and local heating
Cascade impactor	Impaction stages in series collect successively smaller particles on basis of aerodynamic size	Cutoffs for stages can be inprecise. Particle bounce and re-entrainment produce biases; wall losses can be appreciable, and low pressures can cause evaporation of liquids; lower cutoff at about 0.3 μm aerodynamic diameter, except in low pressure models; amount of sample collected is usually small before overloading occurs; particle charges can produce artifacts
Condensation nuclei counters	Supersaturated vapor causes particle growth; particle number determined by intensity reduction of a light beam of air.	Lower limit of detection near 0.002 μm; coincidence errors occur at high concentrations, about 10^6 or 10^7/cc; calibration is difficult
Centrifuge spectrometers	Rotation produces increased sedimentation velocities across a clean air field where particles settle onto a tape	Inlet losses can be large for particles above a few micrometers
Diffusion batteries	Aerosol is drawn through fine screens or small channels in parellel at various air flow rates; measurements of penetrating fractions lead to size distribution	Inappropriate for particles greater than a few tenths of a micrometer in diameter; analysis is increasingly difficult as geometric standard deviation increases
Charge spectrometers	Aerosol is drawn between oppositely charged plates and deposits are analyzed to yield number of charges on particles	Operational difficulties include maintaining stable laminar air flow and prevention of arcing; analysis of polydisperse aerosols is difficult; usually custom designed
Mobility analyzers	Particles are given known charge and selectively deposited with respect to size by change in intensity of a collecting electrical field	Instruments have large internal losses that are particle-size dependent; calibration is sensitive to many environmental factors; lower size limit is undefined; upper size limit is about 1 μm

TABLE 4.8 (continued)
Aerosol Measuring Instrumentation for Inhalation Studies

Instrument	Principle	Artifacts and limitations
Surface area measurement devices	Particles are degassed and then allowed to adsorb gas on their surfaces; amount of adsorbed gas is determined gravimetrically or by other means	Requires large quantities of collected aerosol; method is sensitive to surface character, presence of cracks, and voids in particles; low pressures and high temperatures used may alter particle surface area during measurement
Elutriators	Particles travel usually horizontally and settle through clean air onto a surface; position on the surface is related to aerodynamic diameter	Impractical for submicrometer particles due to their large diffusion coefficients; charged particles may not deposit at proper locations

Source: Phalen, R. F., *Inhalation Studies: Foundations and Techniques,* CRC Press, Boca Raton, FL, 1984, 112–113.

one must accurately determine the collected mass on each of the impactor collection stages, as well as on the backup filter. In some cases, a gravimetric determination is adequate, but often a chemical analysis provides much more specific information. Obviously, one must be sure that the entire mass collected on each stage is recovered and quantitatively assayed. Proper use of impactors requires skill and understanding. If the aerosol is liquid and the impaction surface hard, for example, one may have problems with droplets breaking up and fractionating during collection. On the other hand, if the impactor collection surface is a filter or has been coated with oil or grease to prevent particle bounce, elaborate means may be necessary to remove all of the collected deposit for subsequent analysis. Although the median aerodynamic size can be obtained using a cascade impactor, effects such as particle bounce and electrostatic repulsion will tend to artificially broaden the estimated size distribution.

Many sizing techniques, such as electron microscopy and cascade impaction, have a major shortcoming: they do not usually give a final answer rapidly. Often, many hours or even days will pass before one has a reliable result. Thus, these techniques generally are not useful for real-time adjustments of aerosol generators. To solve this problem, instruments that give real-time responses may also be required. As purchased, such instruments are calibrated against relatively ideal particles (for example, monodisperse polystyrene latex microspheres with densities near 1 g cm^{-3}). Several instruments are available that provide real-time aerosol size distribution data (Table 4.8). In most cases it is necessary to periodically check their accuracy under the specific conditions of an experiment, since instrument calibrations made using more ideal particles may not apply well to other aerosols.

Examples of commonly used real-time monitors include optical particle spectrometers, optical particle counters, condensation nucleus counters, low-pressure piezoelectric impactors, and mobility analyzers. With adequate calibration, real-time monitors can be very useful for both controlling and characterizing atmospheres in inhalation studies.

In some special studies, additional aerosol instrumentation will be required. For example, one may need information on the surface areas or electrical charge states of particles. Thus, specialized instrumentation will be necessary. In such cases, experienced scientific personnel and a thorough knowledge of the aerosol literature are essential for success.

The air-flow patterns entering sampling devices can be complex and lead to collection of a nonrepresentative sample. Mercer[1] has described conditions under which particles of various sizes may be sampled isokinetically. It is very difficult to isokinetically sample particles with diameters of 20 μm or greater, especially when sampler inlets are tubular and sampling flow rates are less than about 1000 l sec^{-1}. Mercer,[1] Hinds,[4] or Vincent[24] should be consulted if large particles are to be sampled.

E. Gas Characterization

Three primary methods are used for obtaining samples of gaseous materials for analysis: grab sampling, absorption sampling, and continuous sampling. In each method, the objective is the same — to acquire a representative sample, that is not altered by the sampling process, for subsequent analysis. The collection efficiency of a gas sampler or sampling procedure must be known in order to quantitatively determine an airborne concentration. The collection efficiency, C, is

$$C = \frac{\text{amount collected or detected}}{\text{amount present in sampled volume of air}} \tag{4.2}$$

The collection efficiency can be less than 1 due to factors such as absorption or adsorption in sampling lines, losses due to chemical reactions, and condensation of gases onto the surfaces of airborne particles. Estimation of the collection efficiency is typically accomplished by measuring concentrations in standardized calibration mixtures. Note that the entire sampling system should be tested in the configuration to be used in the actual study. Great care must be taken in selecting appropriate sampling line materials and in removing particles from the sampled air flow.

Grab sampling refers to the rapid acquisition of an air sample that is then transported to analytical equipment. Evacuated containers, calibrated syringes, plastic or aluminized bags, and a variety of other containers are used to hold grab samples. Problems encountered include: porosity of the container, sample contamination by container materials or contaminant air residing in the container, and degradation of the sample within the container. Adsorption sampling refers to the capture of either a sample on a high surface substrate such as activated charcoal or in a liquid solvent such as water. The collected sample is then driven from the capture medium for analysis or is analyzed *in situ* on the medium. A related type of sampler, the cold trap, collects a gas by low-temperature liquefaction. Several types of adsorption and absorption media are in use, including glass beads (wet, dry, chilled, etc.), activated charcoal, silica gel, ion-exchange resins, impregnated filters, and a variety of other specialized materials.

The analysis of a sample of gas may be performed by traditional quantitative wet-chemistry followed by gravimetric, photometric, or some other analytical method. Other phenomena or analytical techniques that have been exploited for analysis of gases include chemiluminescence (chemically stimulated light emission), electrical conductivity of a solution, heat of combustion, thermal conductivity, coulometry (measurement of charge generated by chemical reactions), flame ionization, gas chromatography, spectrophotometry, polarography (electrolysis), radioactivity, and mass spectrometry. Table 4.9 lists some gases and representative analytical techniques applied to them. Methods for many of the commonly encountered contaminants of ambient air and workplace atmospheres have been published.[25-26] The current trend in developing real time instrumentation for gases includes improvements in specificity, portability, and sensitivity. Reading the current scientific literature is essential to the proper selection of gas-monitoring instrumentation. Useful journals in this regard include: *Journal of the Air & Waste Management Association*, *Environmental Health Perspectives*, *Analytical Chemistry*, *Atmospheric Environment*, *Environmental Science & Technology*, *American Industrial Hygiene Association Journal*, and *Applied Occupational & Environmental Hygiene*.

V. Eliminating Measurement Interferences

Interference occurs when the analysis for one material is significantly altered by the presence of another. In an air sample taken during inhalation exposure, several unwanted materials are often present. Three basic techniques can be applied to the problem of interferences: selection of detection methods that do not suffer from significant interferences, removal of interfering species from the air stream, and correction of analyzer readings to remove that portion of the reading due to interfering species.

Selection of detectors that do not suffer from significant interferences has become easier in recent years. In general, various absorption and emission spectra are unique to each given material, and monitoring instruments based on such spectra tend to be relatively free of interferences. One problem that

TABLE 4.9
Representative Methods for Collection and Analysis of Gases and Vapors

Chemical techniques			
Gas or vapor	Sorption medium	Analysis	Interferences
Ammonia	25 m*l* 0.1 *N* Sulfuric acid	Nessler reagent	—
Benzene	Activated charcoal	Gas chromatography	Other similar retention compounds
Formaldehyde	1% 10 m*l* Sodium bisulfate	Color of product	Phenols, some other aldehydes
Nitrogen dioxide	20–30 m*l* Saltzman reagent	Reacts with adsorbing solution	Ozone in fivefold excess; peroxyacyl nitrate
Ozone	1% Potassium iodide in 1 *N* potassium hydroxide	Measure color of iodine liberated	Other oxidizing agents
Styrene	15 m*l* Spectrograde isoctane	UV analysis	Other aromatic hydrocarbons

Direct reading instruments			
Operating principle	Applications and remarks	Range	Sensitivity
Chemiluminescence	Measurement of NO in ambient air selectively and NO_2 after conversion to NO by hot catalyst; specific measurement of O_3; no atmospheric interferences	0 to 10,000 ppm	Varies: 0.1 ppb to 0.1 ppm
Colorimetry	Measurement and separate recording of NO_2–NO_x, SO_2 total oxidants, H_2S, HF, NH_3, Cl_2, and aldehydes in ambient air	ppb and ppm	0.01 ppm (NO_2, SO_2)
Coulometry	Continuous monitoring of NO, NO_2, O_3, and SO_2 in ambient air; provided with strip chart recorders; some require attention only once a month	Selective: 0 to 1.0 ppm overall or to 100 ppm (optional)	Varies: 4 to 100 ppb dependent on instrument range setting
Infrared analyzer	Continuous determination of a given component in a gaseous or liquid stream by measuring amount of infrared energy absorbed by component of interest using pressure sensor technique; wide variety of applications including CO, CO_2, freons, hydrocarbons, nitrous oxide, NH_3, SO_2, and water vapor	From ppm to 100% depending on application	0.5% of full scale

Source: Phalen, R. F., *Inhalation Studies: Foundations and Techniques*, CRC Press, Boca Raton, FL, 1984, 116.

arises when airborne gases are analyzed spectroscopically is that the presence of particles can lead to unwanted scattering and absorption of light and other electromagnetic radiation. Such interfering particles may be present in the breathing zone even when they are not intentionally generated. Also, particles may be formed within a sampling line or an analytical instrument due to changes in pressure, temperature, humidity, or other factors.

Removal of interfering species from a sample air stream may be necessary in order to obtain valid readings with analytical instruments. Particulate filters can be placed in gas sampling lines. Several problems are associated with such in-line particle filters, including: reaction or capture of the sampled gas on the filter material or particles trapped in and on the filter, reduction of the sampler air-flow rate due to the resistance of the filter, dilution of the sample by air leaks that occur in the filter holder, and introduction of new materials into the sample due to chemical reactions or physical desorption of materials trapped on the filter. Intentional removal of gases from the sampled air can be achieved by coating the walls of a sampling line with specific absorbers or by passage of the sample through a bed or filter containing an absorber. One must be careful in designing such scrubbers to ensure that they do not retain the sampled species.

If it is necessary to correct instrument readings for interfering materials, two general methods can be considered. One may monitor the concentration of the interfering material separately and correct the output of the primary instrument using data on the relationship of the concentration of the interfering substance and the size of the interfering signal. Alternately, one may experimentally recalibrate the primary instrument under actual exposure conditions with interfering species present. The first technique is usually superior when the interfering material has a relatively constant concentration. In order to have confidence in one's ability to correct for or eliminate interferences, it is wise to intentionally generate known amounts of interfering species and thus test the correction technique before using it in an actual inhalation study.

Stevens et al.[27] and Kleinman et al.[28] used an ammonia gas denuder to remove ammonia from sampled air to prevent neutralization of acid particles on a sampling filter. The device described by Stevens consisted of a set of 16 parallel glass tubes (30-cm length, 0.5-cm i.d.) coated with phosphorous acid. A gas diffusion denuder for removal of sulfur dioxide and/or hydrogen sulfide was described by Coburn et al.[29] that had a gas capture efficiency of nearly 100% and permitted 98% or more penetration of particles. Recent denuder designs can be used in tandem to separate and analyze acid gases, basic gases, and particulate species in a single sample.[30-31] Annular denuders have been used very successfully for inhalation studies and environmental studies.[32-36]

VI. Sampling Protocols

Ideally, all of the relevant atmospheric characteristics would be continuously monitored throughout an inhalation exposure. In practice, this is likely to be an impossibility for several reasons. Many samplers do not provide continuous data, and samples must be collected over finite periods in order to accumulate sufficient material for analysis. The total air-flow rate through a large battery of samplers can exceed the total exposure system throughput air-flow rate. Also, the instrument costs and manpower requirements for an ideal, complete monitoring scheme can be prohibitive. Therefore, practicality, ingenuity, and scientific insight are all elements in designing valid sampling protocols that are cost effective.

At a minimum, the primary experimental materials being investigated, as well as any other materials or parameters that might reasonably be expected to modify the measured responses, must be measured. The mean concentration is usually the most important datum relating to the potential biological effect of a material, but the standard deviation, range, and peak values may also be important to interpreting the observed responses. In some studies, the rate of increase or decrease in concentration of a material may be measured, as such phenomena can also influence biologic responses. When a continuous data record is made, each of the foregoing exposure parameters can, in principle, be obtained from the record. When discrete samples are obtained in order to determine a desired mean concentration value, the minimum number of samples (N) that must be taken can be calculated given two additional input parameters: the required precision with which one must know the mean (i.e., the standard error of the mean, SEM) and the expected standard deviation (SD) of the measured concentrations. Then:

$$N = (SD/SEM)^2 \tag{4.3}$$

This equation is based on the assumptions that the concentration measurement data are randomly distributed about the mean (normally distributed) and that each measurement is independent of the others.

When discrete samples are acquired, one cannot determine the maximum and minimum concentrations, as they may have occurred between sample periods or may have been masked by the finite duration of sampling. When discrete samples are acquired, one should thus report not only the number of samples taken, but also the duration of each sampling cycle.

When the time required to acquire a sample for analysis is large with respect to the total duration of the inhalation exposure, one should consider beginning the sample at the start of the exposure and continuing the sampling throughout the exposure period. In this way, the sample will represent an average which is representative of the conditions that existed throughout the exposure.

VII. Suppliers

A. Aerosol Generators

In-Tox Products
115 Quincy NE
Albuquerque, NM 87108

TSI, Inc.
500 Cardigan Road
Shoreview, MN 55126

BGI, Inc.
58 Guinan Street
Waltham, MA 02154

B. Aerosol Sizers and Samplers

In-Tox Products
115 Quincy NE
Albuquerque, NM 87108

SKC, Inc.
863 Valley View Road
Eighty Four, PA 15330

MIE, Inc.
1 Federal Street, #2
Billerica, MA 01821-3500

TSI, Inc.
500 Cardigan Road
Shoreview, MN 55126

Met One, Inc.
481 California Avenue
Grants Pass, OR 97526

Malvern Instruments, Inc.
10 Southville Road
Southborough, MA 01772

Graseby (Andersen Samplers)
500 Technology Court
Smyrna, GA 30082

Climet Instruments Co.
1320 Colton Avenue
Redlands, CA 92374

Particle Measuring Systems, Inc.
5475 Airport Blvd.
Boulder, CO 80301-2239

Quantichrome Corp.
(Surface Area Analysis)
1900 Corporate Drive
Boynton Beach, FL 33426

C. Gas Generators

Liquid Carbonic
Cylinder Gas Products
810 Jorie Blvd.
Oak Brook, IL 60521-2216

Altech Associates, Inc.
2051 Waukegan Road
Deerfield, IL 60015

Matheson Gas Products
30 Seaview Drive
Secaucus, NJ 07094

Scott Specialty Gases
6141 Easton Road
Plumsteadville, PA 18949-0310

D. Gas Analyzers and Samplers

Dasibi Environmental Corp.
506 Paula Avenue
Glendale, CA 91201

SKC, Inc.
863 Valley View Road
Eighty Four, PA 15330

Monitor Labs, Inc.
74 Inverness Drive E.
Eaglewood, CO 80112

Columbia Scientific Industries Corp.
11950 Jollyville Road
Austin, TX 78759

References

1. Mercer, T. T., *Aerosol Technology in Hazard Evaluation,* Academic Press, New York, 1973.
2. Liu, B. Y. H., Eds., *Fine Particles,* Academic Press, New York, 1976.
3. Willeke, K. and Baron, P. A., Eds., *Aerosol Measurement,* Van Nostrand Reinhold, New York, 1993.
4. Hinds, W. C., *Aerosol Technology,* John Wiley & Sons, New York, 1982.
5. Phalen, R. F., *Inhalation Studies: Foundations and Techniques,* CRC Press, Boca Raton, FL, 1984.
6. McClellan, R. O. and Henderson, R. F., Eds., *Concepts in Inhalation Toxicology,* Taylor and Francis, Washington, D.C., 1995.
7. Sinclair, D. and LaMer, V. K., Light scattering as a measure of particle size in aerosols, *Chem. Rev.,* 44, 245, 1949.
8. Prodi, V., A condensation aerosol generator for solid monodisperse particles, in *Assessment of Airborne Particles,* Mercer, T. T., Morrow, P.E., and Stober, W., Eds., Charles C Thomas, Springfield, IL, 1972, chap. 9.
9. Walton, W. H. and Prewett, W. C., The production of sprays and mists of uniform drop size by means of spinning disc type sprayers, *Proc. Phys. Soc.,* 62, 341, 1949.
10. Lippmann, M. and Albert, R. E., A compact electric-motor driven spinning disc aerosol generator, *Am. Ind. Hyg. Assoc. J.,* 28, 501, 1967.
11. Mori, H., Marijnissen, J., and Scarlett, B. Optimal conditions for the generation of monodisperse aerosol particles using the spinning top generator, *J. Aerosol Sci.,* 21, S661, 1990.
12. Raabe, O. G., The dilution of monodisperse suspensions for aerosolization, *Am. Ind. Hyg. Assoc. J.,* 29, 439, 1968.

13. Japuntich, D. A., Stenhouse, J. I. T., and Liu, B. Y. H., An aerosol generator for high concentrations of 0.5 — 5 μm solid particles of practical monodispersity, *Aerosol Sci. Technol.,* 16, 246, 1992.
14. Fulwyler, M. J., Electrical separation of biological cells by volume, *Science,* 150, 910, 1965.
15. Raabe, O. G. and Newton, G. J., Development of techniques for generating monodisperse aerosols with the Fulwyler droplet generator, in *Fission Product Inhalation Program Annual Report, 1969–1970,* LF-43, Lovelace Foundation, Albuquerque, NM, 1970, 13.
16. Strom, L., The generation of monodisperse aerosols by means of a disintegrated jet of liquid, *Rev. Sci. Instrum.,* 40, 778, 1969.
17. Berglund, R. N. and Liu, B. Y. H., Generation of monodisperse aerosol standards, *Environ. Sci. Technol.,* 7, 147, 1973.
18. Mitchell, J. P., Snelling, K. W., and Stone, R. L. Improvements to the vibrating orifice aerosol generator, *J. Aerosol Sci.,* 18, 231, 1987.
19. Liu, B. Y. H. and Pui, D. Y. H., Electrical neutralization of aerosols, *J. Aerosol Sci.,* 5, 465, 1974.
20. Marple, V. A., Lui, B. Y. H., and Rubow, K. L., A dust generator for laboratory use, *Am. Ind. Hyg. Assoc. J.,* 39, 26, 1978.
21. Lua, A. C., Particle characteristics of a stable fluidized bed aerosol generator, *J. Aerosol Sci.,* 23,737, 1992.
22. Wright, B. M., A new dust-feed mechanism, *J. Sci. Instrum.,* 27, 12, 1950.
23. Lodge, J. P. and Chan, T. L., Eds., *Cascade Impactor: Sampling and Data Analysis,* American Industrial Hygiene Association, Akron, OH, 1986.
24. Vincent, J. H., *Aerosol Science for Industrial Hygienists,* Pergamon Press, Elmsford, NY, 1995.
25. Lodge, J. P. Jr., Ed., *Methods of Air Sampling and Analysis,* Third ed., Lewis Publishers, Boca Raton, FL, 1989.
26. Eller, P. C. and Cassinelli, M. E., *NIOSH Manual of Analytical Methods,* Fourth ed., U.S. Department of Health and Human Services, Public Health Service, Centers for Disease Control, Cincinnati, OH, 1994.
27. Stevens, R. K., Dzubay, T. G., Russwurm, G., and Rickel, D., Sampling and analysis of atmospheric sulfates and related species, *Atmos. Environ.,* 12, 55, 1978.
28. Kleinman, M. T., Bailey, R. M., Chang, Y. T. C., Clark, K. W., Jones, M. P., Linn, W. S., and Hackney, J. D., Exposures of human volunteers to a controlled atmospheric mixture of ozone, sulfur dioxide and sulfuric acid, *Am. Ind. Hyg. Assoc. J.,* 42, 61, 1981.
29. Coburn, W. G., Husar, R. B., and Husar, J. D., Continuous in situ monitoring of ambient particulate sulfur using flame photometry and thermal analysis, *Atmos. Environ.,* 12, 89, 1978.
30. Koutrakis, P., Wolfson, J. M., Brauer, M., and Spengler, J. D., Design of a glass impactor for an annular denuder/filter pack system, *Aerosol Sci. Technol.,* 12, 607, 1990.

31. Thomas, C. L. P. and Alder, J., Annular denuder tube preconcentrator for nitrobenzene determination by gas chromatography with electron-capture detection, *Anal. Chim. Acta,* 217, 289, 1989.
32. Perrino, C., DeSantis, F., and Febo, A., Criteria for the choice of a denuder sampling technique devoted to the measurement of atmospheric nitrous and nitric acids, *Atmos. Environ.,* 24A, 617, 1990.
33. Waldman, J. M., Lioy, P. J., Thurston, G. D., and Lippmann, M., Spatial and temporal patterns in summertime sulfate aerosol acidity and neutralization within a metropolitan area, *Atmos. Environ.,* 24B, 115, 1990.
34. Koutrakis, P., Wolfson, J. M., and Spengler, J. D., An improved method for measuring aerosol strong acidity: results from a nine-month study in St. Louis, Missouri, and Kingston, Tennessee, *Atmos. Environ.,* 22, 157, 1988.
35. Grosjean, D., *In situ* organic aerosol formation during a smog episode: estimated production and chemical functionality, *Atmos. Environ.,* 26A, 953, 1992.
36. Williams, E. L., II, and Grosjean, D., Removal of atmospheric oxidants with annular denuders, *Environ. Sci. Technol.,* 24, 811, 1990.

Chapter 5

Inhalation Exposure Methods

Robert F. Phalen

Contents

0-8493-3343-1/97/$0.00+$.50

I. Introduction

A. Goals

When selecting an inhalation exposure system, the basic requirement is that it deliver a stable, well characterized study atmosphere to a subject or subjects for a specified duration. Inhalation exposure durations vary from a few seconds to a major fraction of the lifespans of the subjects (months or years). Secondary, but important, goals are to minimize stress to the subjects, to prevent exposure to substances other than those under study, to protect laboratory personnel from exposure to the test substances, and to prevent or limit nonrespiratory tract exposure (dermal and gastrointestinal) of the study subjects. In addition, a particular study may have additional objectives, such as permitting exercise or varying the exposure concentration on a predetermined schedule. Achieving all of the above goals simultaneously may be challenging, even for an experienced investigator and team. Several excellent papers and general reference works describing exposure systems that have been successful in meeting these goals are available.[1-23]

B. Systems Available

Over the past 50 years, a large variety of sophisticated inhalation exposure systems has been designed and used for exposing human and animal subjects to various test agents. A useful means of classifying these inhalation exposure systems is to consider how much of the subject's body is in direct contact with the atmosphere during an exposure (Table 5.1). Whole-body exposure (or immersion) systems employ chambers in which the subjects can move about, either unrestrained or in cages. Such systems permit exposure of large numbers of subjects, are relatively nonstressful, are efficient with respect to the technician labor involved, and are good for chronic exposures where daily exposure durations are 6 to 24 hours. On the other hand, such systems expose all surfaces of the subjects and often their food and water; they can require large volumes of study material and throughput air; they are expensive to build, install, and maintain; and the exposure atmosphere is often heavily contaminated by dander, excreta, microorganisms, and ammonia.

Head-only exposure systems are of two basic types: helmets and portals that permit insertion of the head into a chamber. These systems work well for brief, repeated exposures (because food and water usually are not able to be provided), prevent much of the dermal exposure, require less study material than do chambers, and allow one to quickly vary the exposure concentration

TABLE 5.1
The Major Advantages and Disadvantages Associated with Various Modes of Inhalation Exposure

Mode	Advantages	Disadvantages
Chamber	• Large number of subjects • Suitable for chronic studies • Minimal restraint • Can house animals in chambers • Labor efficient	• Dermal, eye, and oral exposure in addition to inhalation • Large amounts of test material required • Expensive • Excreta can interact with pollutants
Head-only	• Good for repeated exposure • Minimal skin contamination • Efficient dose delivery • Better control of dose	• Can be stressful • Pollutant losses can be large • Neck seal problems • Labor intensive
Nose/mouth-only	• No skin contamination • Can be used for repeated exposures • Uses much less material • Exposures can be pulsed • Personnel and facility contamination minimized	• Can be stressful • Needs good face seal • Labor intensive • Technically difficult
Lung-only	• Precision of dose • Uses less material (efficient)	• Anesthesia or tracheostomy required • Bypasses nose (could be an advantage) • Artifacts in deposition and response • Technically difficult

due to the air volume being much smaller than that needed for a whole-body chamber. Disadvantages associated with these systems include the possibility of producing stressful restraint for the subjects, the possibility of losing much of the study material to helmet surfaces, difficulty in maintaining a closed but comfortable neck seal, and the additional labor involved in positioning subjects and repeatedly checking neck seals.

Nose-only or mouth-only exposure systems include tapered tubes (for rodents and other burrowing animals), masks (for humans, dogs, rabbits, etc.), and oral-cavity apparatuses that position an exposure system tube inside the mouth (these often require full anesthesia to prevent stress or chewing on the tube). The disadvantages associated with these systems are similar to those of head-only systems, plus they are technically more demanding because they require that the system be engineered to fit the subject with more precision. Despite these problems, the advantages of nose-only exposures are considerable. They totally eliminate nonrespiratory pathways of exposure, are very efficient with respect to the volumes of study material and throughput air that are needed, eliminate contamination of the air with dander and excreta, and the control of the study material can be very precise. Because of the success and versatility of nose-only exposure systems and the availability of tried-and-true designs, this chapter will emphasize their use.

Exposure systems that expose only portions of the respiratory tract have also been used when precise delivery of a material to a limited area of tissue is required. Full anesthesia is required, frequently along with surgical implanting techniques. These systems tend to be nonphysiological in their delivery of test agents and are used only in specialized circumstances. The use of techniques other than inhalation, such as instillation of liquids, is frequently criticized on the basis of the trauma produced locally in airways; however, any method may be very useful for specific research purposes.

C. Ethical Considerations

Ethical considerations and good scientific practice require that the investigator have an intimate knowledge of stresses placed on the animals. The ethical challenges involved in conducting inhalation studies are not trivial. In an inhalation experiment, the subject is critically dependent on the exposure system for delivery of a breathable atmosphere and a comfortable environment. Without adequate monitoring systems, animals can be placed in situations with high air concentrations of ammonia and carbon dioxide, extremes in humidity and temperature, and uncomfortable confinement. Inexperienced investigators may have no awareness of these stressors. Poorly fitted masks, collars, slings, and other restraint devices can produce great stress. Padding, sedation, or anaesthesia are necessary in many cases. Horses, donkeys, pigs, and some other animals require freedom of movement of the head when they are restrained in an unanesthetized state for more than brief periods. Animals can overheat in close-fitting enclosures unless provision is made for cooling. Rats cool via their tails, so they are more stable in systems that do not confine their tails. Training the animals by putting them through trial sham exposures ("dry runs") can help them to relax and thus improve their performance during actual exposures. A good rule is to only use an exposure system in which you would be willing to be exposed. Also, laboratory personnel should be trained to respect and properly handle animals, as well as to recognize distress and to quickly alleviate it.

D. Safety

Inhalation studies pose some relatively serious safety problems. In addition to the safety considerations that are associated with the use of any major laboratory equipment, exposure systems have special requirements. The study material may be hazardous, so seals should be checked, the atmosphere should be under slight negative pressure in relation to the surroundings, and some means of detecting the contamination of laboratory air or personnel should be considered in the study design. Other safety considerations relate to the study subjects themselves. Rodents are notorious for generating airborne allergens that can sensitize laboratory personnel, making their tasks unpleasant or, in some cases, nearly impossible to perform. In addition, many species of laboratory animals will bite when they are tired, uncomfortable, disoriented, agitated, handled roughly, or, for a few species and strains, whenever the opportunity presents. Bites are prevented by gentle, proper handling; eliminating stress; acclimating the animals to the procedures; and using soft gloves. Only if absolutely necessary should one use chain-mail gloves with small animals (mice and hamsters), since such gloves are sometimes associated with producing tissue damage in small animals. Anticipating safety hazards and applying preventative measures, as opposed to responding to preventable emergencies, are signs of an experienced inhalation toxicologist.

II. Problems Encountered

A. Materials

Exposure systems are usually constructed of relatively inert materials such as metals, glass, and plastics. Problems arise when the study material is chemically reactive. For example, acidic aerosols and several nonacidic salts can corrode many metals or even react to produce hazardous byproducts. Various grades of stainless steel are available and often are used, but they can greatly increase the expense of the system. Nonconductive materials (glass, rubber, and plastic) can easily acquire electrostatic charges that collect study aerosols. This problem can be severe enough to nearly totally deplete the atmosphere of particulate prior to it being inhaled. Antistatic coatings can be applied, and some investigators dampen the surfaces of their systems with aqueous detergent, followed by air-drying (no touching or rubbing with cloth or paper), in order to destroy islands of electrical charge. The aerosol particles can also be discharged by passage through an ion field prior to use.[9]

B. Pollutant Mixing

A poorly designed exposure system can have poor mixing characteristics. That is, the study material is not uniformly dispersed in the throughput air, producing a uneven exposure of the subjects. Careful attention must be paid to how the study material is initially dispersed into the air. Use of venturi mixers (in which the pollutant/air mixture is contracted and then expanded), turbulent mixers, and multiple injection ports (often in a radial design) have been used; however, the uniformity of exposure must be verified in each case.

C. Contaminants

Even if the air supply systems are highly filtered and scrubbed of gaseous pollutants, unwanted contaminants can be present at the breathing zones of the subjects. Such contaminants arise from the air-moving and air-conditioning equipment and from the subjects themselves. Animal dander can add a substantial quantity of unwanted contaminant particles to the breathing zone of the exposure system. A severe problem arises when bacteria in the excreta generate ammonia.[24] State-of-the-art exposure systems are meticulously maintained and cleaned so that excreta-generated ammonia does not build up. In some cases, the use of antibiotics, disinfectants, or other substances where excreta accumulate can be used to suppresses ammonia. Rock salt can be used below wire cage bottoms to dry feces and prevent ammonia formation. Control of animal-generated contaminants is also discussed in Chapter 2.

D. Sampling

Samplers must be used to measure the quantities of test agents in the study atmosphere in order to characterize and control the exposure. Samplers should be properly calibrated and extract representative air samples from the breathing zones of the subjects. Otherwise, the samples may not accurately represent what is inhaled. Also, significant losses of the study material in sampling lines must be prevented. This is accomplished by using large-bore, short length, chemically inert, and electrostatically uncharged lines for sampling aerosols. Metals or conductive plastic lines will not acquire local charges that collect the study aerosol before it can be analyzed. If nonconductive materials are used, one should consider using a static-elimination procedure to treat lines and surfaces prior to the study. In cases in which reactive gases (such as ozone, nitrogen dioxide, etc.) or vapors (such as formaldehyde, etc.) are sampled, it is wise to use inert fluorocarbon tubing for sampling lines.

E. Animal Handling

Each species of laboratory animal must be handled in a unique manner to prevent stress and to promote cooperation. Gentle handling and taking care to provide secure support (to prevent dropping the animal) are essential. As a precaution, the fingers of those handling animals should not be placed in front of an animal's muzzle, as fingers can be mistaken for food, resulting in a nip or even a severe bite. Experienced animal handlers should be used to train laboratory personnel in the proper techniques. Also, gentleness in off-study daily care is essential to helping animals relax and willingly accept handling during a study.

III. A Custom-Made, Nose-Only Exposure System

Some excellent commercially available nose-only exposure systems are available (see suppliers of equipment), but it is feasible to convert a whole-body chamber into a nose-only exposure system that has excellent performance. Smith et al.[25] gave a description of such a system that was used at the Los Alamos National Laboratory for exposing rodents to mineral fibers. The authors replaced their chamber doors with two panels, each with 32 or 45 ports, to receive tapered plastic tubes with one end open for nose-only exposure of rats or hamsters. As a result of the success of this conversion, the Air Pollution Health Effects Laboratory (University of California, Irvine) converted four University of Rochester-type (1-m^3 stainless steel) chambers to permit nose-only exposures of rats.[26] In this system (Figure 5.1), which is still used, all six walls of each hexagonal chamber were modified to expose 20 rats nose-only, giving a total capacity for all four chambers of 480 rats. Inside the chambers, each nose-only port is independently supplied with exposure atmosphere by a rectangular cross-section stainless steel channel. The air flow through these channels is adjusted to be greatly in excess of each rat's ventilation requirement to prevent rebreathing of the flowing atmosphere. On the outside of each wall, short cylindrical stainless steel tubes were welded such that they lead to each internal channel. These tubes accept plastic body tubes with O-ring-fitted aluminum nose-pieces that were designed to accept the rat's heads, permitting about 1 cm of the snout to protrude into the flowing exposure atmosphere. A plastic tailgate at the rear of the body tube is dropped into one of three slots at different distances from the front of the tube, thus preventing the rats from backing out of the exposure tubes. The tailgate covers about 3/4 of the rear opening, so that the rat's tail can hang outside of the tube (rats cool themselves via blood-flow to the tail).

Early in the use of this system, exposure technicians started to experience a moderately severe inflammatory response to the airborne dander and dried excreta from the animals. To solve this problem, external box-like enclosures were installed around the exposure tubes and a negative pressure was applied to scavenge the animal-generated dust and to ventilate (cool) the enclosures. In the intervening years, this system has been used very successfully to expose thousands of animals to a variety of particles, gases, and particle-gas mixtures. Technical personnel like the system, and rats accept the exposure system well for 4 hours/day, 5 days/week, for up to 9 months. This system is a good alternative to the commercial nose-only exposure modules.

IV. Commercial Nose-Only Exposure Systems

A. Historical Development

Turning now to commercially available, nose-only exposure systems, it is useful to examine each component of the systems in some detail. A system in

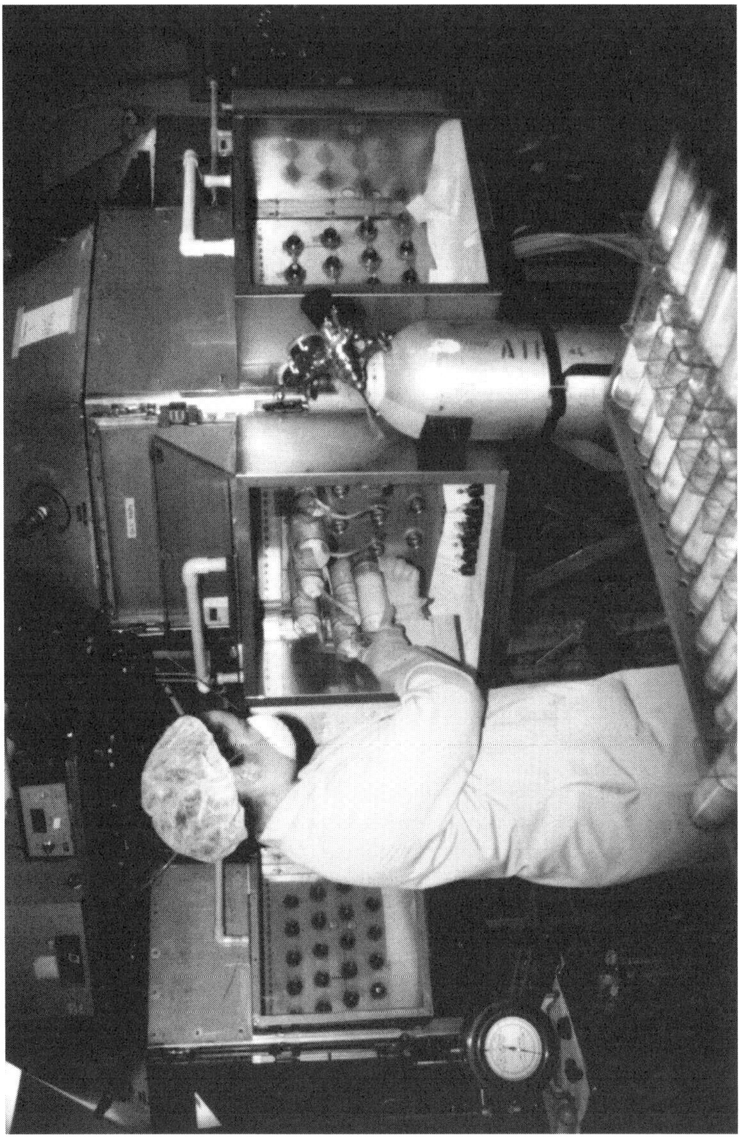

Figure 5.1
Nose-only exposure system made by modifying a Rochester-type, whole-body inhalation exposure chamber.

use at the Air Pollution Health Effects Laboratory (University of California, Irvine) can be traced to an original unit designed for use at the Inhalation Toxicology Research Institute (ITRI; Albuquerque, NM) by Raabe et al.[27] The original unit had a tapered aerosol entry section on top to present a mixed, uniform concentration of particles to the main chamber below. Two walls of the rectangular main chamber each had 44 (for rats, 88 for hamsters) brass ports into which plastic nose-only exposure tubes were inserted. Like the system described in the previous section, rat noses protrude into the chamber through holes in tapered ends of the tubes (the rear of each tube is closed with a tightly-fitting rubber stopper). Inside the main chamber, a suction system draws exposure atmosphere through individual holes in front of each rat's nose and then into an exhaust manifold. This system has been used successfully for several decades. Improvements in design have been made over the years to decrease the size and cost, while still preserving the excellent containment and uniformity of exposure.

B. The Manifold

Two basic types of nose-only exposure units are commercially available; one has a tower having a circular cross-section (after a design by Cannon et al.),[28] and the other is a narrow rectangular box with exposure ports on two sides (Figure 5.2). Inside the rectangular system, air is drawn through holes to contact each animal's nose and is exhausted through radial holes surrounding each animal's nose. This ensures that each animal gets a fresh, continually renewed supply of the study material with minimal losses of material to the system surfaces. In the tower system, air flows downward past a stack of noses; side ports expose each animal separately. In both types of systems, an airflow is maintained that exceeds the ventilation requirements of the animals. Also, in each system, at least one port is used for obtaining air samples.

V. Exposure Tubes

Various designs of nose-only exposure tubes have been used for exposures. Two successful types (one all plastic and the other a plastic body tube with a machined metal headpiece) both have a tapered internal conical cavity for the animal's head. This taper is important, not only for animal comfort, but it also produces a partial seal around the nose when the animal moves forward to the front of the tube. Animals placed in such tubes usually quickly move to the front as far as possible and stick their noses out of the front hole. If the tubes are too loose around the body, animals may turn around to face the rear of the tube. If the tubes are too tight, animals will not enter them, or will be very restless when they are exposed in them. It is wise to have a collection of various-size tubes available to accommodate unusually small or large animals.

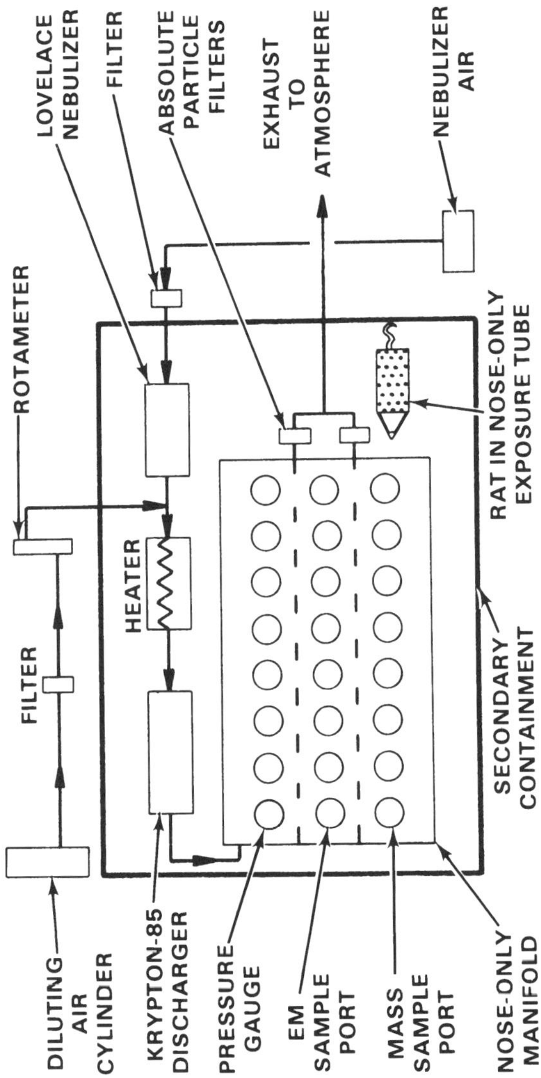

Figure 5.2
Nose-only exposure unit of the type designed by Raabe et al.[27]

Thermal stress can occur due to heat buildup inside nose-only exposure tubes. This heat can produce fatalities if the nose-only system is in a hot surrounding environment. Precautions must be taken to prevent heat stress during confinement. Several methods have proven to be successful, including perforated metal tubes,[29] allowing rat's tails to extend outside of the tubes,[15,26] and keeping the surrounding laboratory air cool.

VI. Air Supply

The air supply system should provide sufficient air to substantially exceed the ventilation requirements of the animals in order to prevent rebreathing of exhaled air. A useful empirical formula for calculating the ventilation per minute (V_m) of a resting mammal of body mass, M (in grams), was published by Guyton:[30]

$$V_m(\text{cm}^3/\text{min.}) = 2.18\ M^{3/4} \tag{5.1}$$

For a 70-kg human, this gives a minute ventilation of 9.4 *l*, and for a 200-g rat it gives a result of 116 cm^3/minute; both values are quite reasonable. To be certain that rebreathing of expired air does not occur, even if the animals become active, it is wise to multiply the value obtained using Guyton's equation by a factor of 10, and then to deliver this volume of air to each subject each minute. As was discussed in Chapter 3, the air supply should be within comfortable limits of temperature and humidity, and it should be free of all significant non-study-related contaminants.

VII. Exhaust Scrubbing

The unfiltered exhaust air from an inhalation exposure system will contain the study material and possibly substances that are added by the subjects (dander, dried particles from feces, saliva, etc.); therefore, scrubbing should be considered. For particulate material, a battery of filters — a coarse filter followed by a high-efficiency filter — is a good solution. For gaseous pollutants, adsorption, absorption, or chemical degradation may be used. Activated charcoal is a good adsorption material, as are several other high-surface-area granular materials. Absorption in a liquid via a bubbler or by passage through scrubber tubes coated with absorbing materials should be considered. Chemical degradation reactions can occur in beds of potassium permanganate impregnated on activated alumina pellets (Purafil, Inc.; Doraville, GA) and in bubblers filled with aqueous acids or bases, for example. Thermal degradation, as by passage of the exhausted air through a heated metal, glass, or ceramic tube, also can be effective. In any case, the investigators must make sure that hazardous or nuisance levels of contaminants are not emitted into the environment.

VIII. Monitoring

In order to specify and control an inhalation exposure, several parameters should be monitored. In relation to the exposure, the study material concentration at the breathing zone and air-flow rate should be known throughout the exposure. If aerosol particles are being studied, the size distribution (preferably including the mass median aerodynamic diameter and geometric standard deviation) should be measured at least once during each daily exposure, preferably more often. (See the general references listed at the beginning of this chapter and in Chapter 4 for acceptable methods.) In high-quality studies, the air temperature and humidity, both inside and surrounding the inhalation exposure system, will be monitored and corrected, if necessary.

It is also sometimes useful to monitor the breathing patterns of animals during exposure. This is very difficult to do if a large number of animals are exposed at once, so monitoring the breathing pattern of one, two, or three randomly selected animals is desirable. See Chapter 6 for a description of the techniques and equipment used for this purpose.

IX. Suppliers of Equipment

A. Chambers

Hazelton Systems
P.O. Box 700
Aberdeen, MD 21001

B. Nose-Only Exposure Systems

CH Technologies (U.S.A.)
263 Center Avenue
Westwood, NJ 07675

In-Tox Products
115 Quincy NE
Albuquerque, NM 87108

Lab Products, Inc.
255 W. Spring Valley Avenue
Maywood, NJ 07607

Unifab Corp.
5260 Lovers Lane
Kalamazoo, MI 49002

Machine Shop (D. Baugh & Son)
(custom-made exposure tubes for small animals)
3413 Fern Meadow Road
Palomar Mountain, CA 92060-0130

X. Sample Data

The major data that are acquired during an inhalation exposure will relate to the characteristics of the study material. Achieving the target concentrations and the target particle-size distributions (if aerosols are involved) are of primary importance. Samples of data from actual inhalation exposures are shown in Table 5.2. This study involved two gases (ozone and nitric acid vapor) that are both often difficult to stabilize. Ozone is very reactive, so system surfaces had to be clean and inert, and the generator was stabilized by use of a voltage regulator. Nitric acid adsorbs onto system surfaces, so preconditioning for several hours was necessary before any animals were placed into the exposure system.

TABLE 5.2
Summary of Atmosphere Characteristics in a Nose-Only Exposure

Exposure atmosphere	Parameter/ pollutant	Target	Measured (mean ± SD)
	4 hours/day × 1 day		
Purified air	Relative humidity	60%	60.7 ± 1.0
	Temperature	22–24°C	23.7 ± 0.8
0.6 ppm O_3	Relative humidity	60%	60.7 ± 1.0
	Temperature	22–24°C	23.7 ± 0.8
	Ozone	0.6 ppm	0.61 ± 0.02
1 mg/m^3 HNO_3	Relative humidity	60%	60.9 ± 0.9
	Temperature	22–24°C	23.3 ± 0.5
	Nitric acid	1.0 mg/m^3	1.08 ± 0.13
0.6 ppm O_3 + 1 mg/m^3 HNO_3	Relative humidity	60%	60.5 ± 1.0
	Temperature	22–24°C	24.1 ± 0.8
	Ozone	0.6 ppm	0.60 ± 0.01
	Nitric acid	1.0 mg/m^3	1.01 ± 0.12
	4 hours/day × 4 days		
Purified air	Relative humidity	60%	60.2 ± 0.9
	Temperature	22–24°C	22.5 ± 0.8
0.15 ppm O_3	Relative humidity	60%	60.2 ± 0.9
	Temperature	22–24°C	22.5 ± 0.8
	Ozone	0.15 ppm	0.15 ± 0.01
0.25 mg/m^3 HNO_3	Relative humidity	60%	60.4 ± 0.9
	Temperature	22–24°C	22.3 ± 0.6
	Nitric acid	0.25 mg/m^3	0.27 ± 0.08

TABLE 5.2 (continued)
Summary of Atmosphere Characteristics in a Nose-Only Exposure

Exposure atmosphere	Parameter/ pollutant	Target	Measured (mean ± SD)
0.15 ppm O_3 +	Relative humidity	60%	60.0 ± 0.8
0.25 mg/m^3 HNO_3	Temperature	22–24°C	22.8 ± 0.8
	Ozone	0.15 ppm	0.15 ± 0.01
	Nitric acid	0.25 mg/m^3	0.26 ± 0.09

Note: No aerosol-phase HNO_3 was observed; all of the HNO_3 was in the vapor phase.

Source: Adapted from Nadziejko, C. E. et al., *Inhal. Toxicol.*, 4, 343, 1992.

References

1. Silver, S. D., Constant flow gassing chambers: principles influencing design and operation, *J. Lab. Clin. Med.*, 31, 1153, 1946.
2. Leach, L. J., Spiegel, C. J., Wilson, R. H., Sylvester, G. E., and Lauterbach, K.E., A multiple chamber exposure unit designed for chronic inhalation studies, *Am. Ind. Hyg. Assoc. J.*, 20, 13, 1959.
3. Frazer, D. A., Bales, R. E., Lippmann, M., and Stokinger, H. E., *Exposure Chambers for Research in Animal Inhalation,* U.S. Public Health Service Publ. No. 57, U.S. Department of Health, Education and Welfare, U.S. Government Printing Office, Washington, D.C., 1959.
4. Hinners, R. G., Burkart, J. K., and Contner, G. L., Animal exposure chambers in air pollution studies, *Arch. Environ. Health*, 13, 609, 1966.
5. Hanna, M. G., Jr., Nettesheim, P., and Gilbert J. R., Eds., *Inhalation Carcinogenesis*, U.S. Atomic Energy Commission, Oak Ridge, TN. Available from the National Technical Information Service, Springfield, VA, as CONF-691001, 1970.
6. Lippmann, M., Experimental inhalation studies — equipment and procedures, in *Inhalation Carcinogenesis,* Hanna, M. G., Jr., Nettesheim, P., and Gilbert, J. R., Eds., U.S. Atomic Energy Commission, Oak Ridge, TN, 1970, 55.
7. Drew, R. T. and Laskin, S., Environmental inhalation chambers, in *Methods of Animal Experimentation,* Gay, W. I., Ed., Academic Press, New York, 1973, chap. 1.
8. Phalen, R. F., Inhalation exposure of animals, *Environ. Health Perspect*, 16, 17, 1976.
9. Phalen, R. F., *Inhalation Studies: Foundations and Techniques,* CRC Press, Boca Raton, FL, 1984, chap. 5.
10. Willeke, K., Ed., *Generation of Aerosols and Facilities for Exposure Experiments,* Ann Arbor Science, Ann Arbor, MI, 1980.
11. Bernstein, D. M. and Drew, R. T., The major parameters affecting temperature inside inhalation chambers, *Am. Ind. Hyg. Assoc. J.*, 41, 420, 1980.

12. Leong, B. K. J., Ed., *Inhalation Toxicology and Technology*, Ann Arbor Science, Ann Arbor, MI, 1981.
13. MacFarland, H. N., Designs and operational characteristics of inhalation exposure equipment — a review, *Fundam. Appl. Toxicol.*, 3, 603, 1983.
14. Salem, H., Ed., *Inhalation Toxicology: Research Methods, Application and Evaluation*, Marcel-Dekker, New York, 1987.
15. Pauluhn, J., Different methods used in acute and subchronic inhalation studies of potential lung irritants, with particular attention to lung function measurements, in *Inhalation Toxicology: The Design and Interpretation of Inhalation Studies and Their Use in Risk Assessment,* Dungworth, D., Kimmerle, G., Lewkowski, J., McClellan, R.O., and Stöber, W., Springer-Verlag, New York, 1988, chap. 6.
16. Dungworth, D., Kimmerle, G., Lewkowski, J., McClellan, R.O., and Stöber, W., Eds., *Inhalation Toxicology: The Design and Interpretation of Inhalation Studies and Their Use in Risk Assessment*, Springer-Verlag, New York, 1988, chaps. 5, 7–8.
17. Moss, O. R., A chamber providing uniform concentration of particulates for exposure of animals on tiers separated by catch pans, in *Inhalation Chamber Technology*, Drew, R. T., Ed., BNL Formal Report No. 51318, Brookhaven National Laboratory, Upton, NY, 1981.
18. Gardner, D.E., Crapo, J.D., and McClellan, R.O., Eds., *Toxicology of the Lung*, Second ed., Raven Press, New York, 1993.
19. Phalen, R. F., Kleinman, M. T., Mautz, W. J., and Drew, R. T., Inhalation exposure methodology, in *Respiratory Toxicology and Risk Assessment*, Jenkins, P. G., Kayser, D., Muhle, H., Rosner, G., and Smith, E. M., Eds., Wissenschaffliche Verlagsgesselschaft mbH, Stuttgart, 1994, 59.
20. Jenkins, P. G., Kayser, D., Muhle, H., Rosner, G., and Smith E. M., Eds., *Respiratory Toxicology and Risk Assessment*, Wissenschaftliche Verlagsgesellschaft mbH, Stuttgart, 1994.
21. Cheng, Y. S. and Moss, O. R., Inhalation exposure systems, in *Concepts in Inhalation Toxicology,* Second ed., McClellan, R.O. and Henderson, R.F., Eds, Taylor and Francis, Washington, D.C., 1995, chap. 2.
22. McClellan, R.O. and Henderson, R. F., Eds., *Concepts in Inhalation Toxicology*, First ed., Hemisphere, New York, 1989.
23. McClellan, R. O. and Henderson, R.F., Eds., *Concepts in Inhalation Toxicology*, Second ed., Taylor and Francis, Washington, D.C., 1995.
24. Barrow, C. S. and Dodd, D. E., Ammonia production in inhalation chambers and its relevance to chlorine inhalation studies, *Toxicol. Appl. Pharmacol.*, 49, 89, 1979.
25. Smith, D. M., Ortiz, L. W., Archuleta, R. F., Spalding, J. F., Tillery, M. I., Ettinger, H. J., and Thomas, R. G., A method for chronic nose-only exposures of laboratory animals to inhaled fibrous aerosols, in *Proceedings of the Inhalation Toxicology and Technology Symposium*, Leong, B. K. J., Ed., Ann Arbor Science, Ann Arbor, MI, 1981, 89.

26. Prasad, S. B., Rao, S. V., Mannix, R. C., and Phalen, R. F., Effects of pollutant atmospheres on surface receptors of pulmonary macrophages, *J. Toxicol. Environ. Health*, 24, 385, 1988.
27. Raabe, O. G., Bennick, J. E., Light, M. E., Hobbs, C. H., Thomas, R. L., and Tillery, M. I., An approved apparatus for acute inhalation exposure of rodents to radioactive aerosols, *Toxicol. Appl. Pharmacol.*, 26, 264, 1973.
28. Cannon, W. C., Blanton, E. F., and McDonald, K. E., The flow-past chamber: an improved nose-only exposure system for rodents, *Am. Ind. Hyg. Assoc. J.*, 44, 923, 1983.
29. Kenoyer, J., Phalen, R., and Davis, J., Particle clearance from the respiratory tract as a test of toxicity: effects of ozone on short and long term clearance, *Exp. Lung Res.*, 2, 111, 1981.
30. Guyton, A. C., Analysis of respiratory patterns in laboratory animals, *Am. J. Physiol.*, 150, 78, 1947.
31. Nadziejko, C. E., Nansen, L., Mannix, R. C., Kleinman, M. T., and Phalen, R. F., Effect of nitric acid vapor on the response to inhaled ozone, *Inhal. Toxicol.*, 4, 343, 1992.

Chapter 6

Animal Monitoring

William J. Mautz

Contents

0-8493-3343-1/97/$0.00+$.50

I. Introduction

A. Objectives

The purpose of a typical experimental inhalation exposure of animals is to deliver a quantifiable dose of airborne material to the respiratory tract under conditions that match the dynamics of natural exposures as closely as possible. There are many factors that influence the inhaled dose of a test material; these include the structure and flow characteristics of the exposure system and the physiological status of the test animals. Characteristics of exposure delivery systems are treated in Chapter 5. The objective of this chapter is to review the respiratory physiological responses that affect inhalation dose-delivery to animals and to describe techniques for monitoring these responses. Animals respond physiologically to the conditions of caging and restraint during exposure, as well as to the inhaled airborne compounds themselves. Monitoring these physiological responses is important, both for characterization of inhaled dose and dose distribution and for measuring biological responses to inhaled substances. However, the demands placed on investigators to make such measurements accurately can be substantial, and training and experience in physiology are usually required.

B. Respiratory Monitoring

The most common physiological variables monitored during inhalation exposures are the breathing rate or respiratory frequency (f, in units of min^{-1}) and the tidal volume (V_T, in m*l* or *l*), which is the volume of air in a breath. The product of breath frequency and tidal volume is the minute ventilation ($\dot{V}_E$, in m*l* min^{-1} or *l* min^{-1}), and this is an index of inhaled dose rate of an airborne compound. By monitoring f and V_T continuously or intermittently during exposure, one can obtain an average $\dot{V}_E$ that is proportional to the inhaled dose rate.

Exercise is sometimes used in exposures of animals to increase the dose rate and to more fully involve the gas exchange capacity of the respiratory system. The respiratory system is designed to provide effective gas exchange over a range of exercise levels, and most mammals can increase gas exchange over resting levels by up to a maximum factor of about 10. Human athletes can increase gas exchange rates up to 16 times resting levels. Exercise not only increases the inhaled dose rate, but it also alters upper airway filtration of inhaled pollutants, which exacerbates the pulmonary effects of many compounds.[1-7] Furthermore, most exposures of people to toxic substances involve some level of physical activity or exercise, particularly in recreational and occupational settings. Human inhalation exposure studies almost universally utilize an exercise protocol, and a few animal studies have also used exercise models.[5-6,8-12] Metabolic rate is commonly used as a measure of workload in exercise studies, and metabolic rate during exposure can be measured as oxygen consumption or carbon dioxide production ($\dot{V}_{O_2}$ or $\dot{V}_{CO_2}$ in m*l* min^{-1} or m*l* min^{-1} kg^{-1} body mass). In addition to serving as a measure of exercise work rate, metabolic gas exchange is often affected by exposure to airborne irritant compounds and may itself be among the important biological endpoints for measuring the effects of toxicant exposure. Changes in metabolic rate may also be reflected by changes in body temperature and heart rate, and minute ventilation will be closely coupled to metabolic gas exchange. Therefore, responses during the course of exposure can profoundly influence the integrated inhaled dose. For example, rats exposed to 0.8 ppm O_3 for 3 hours exhibited reduced $\dot{V}_{O2}$ and $\dot{V}_E$ by up to 20%, compared to purified air exposed controls.[13]

Depression of metabolism, ventilation, and body temperature in response to toxic exposure is commonly observed in small rodents.[14-16] Rodents generally may have a more labile thermoregulatory physiology than other eutherian mammals, and many species have specific adaptations for reducing metabolism in response to varied physiological stresses, such as exposure to cold and to water or nutritional deprivation.[13] Physical restraint is also a potential stressor which, in interaction with cold ambient temperatures, can induce depression of body temperature. Body temperature declines of 2 to 20°C have been observed in restrained rabbits and rodents.[17] Depression of metabolic rate and ventilation serves to reduce inhaled dose rate, and reduced metabolic rate may also mitigate the immediate toxic effects of xenobiotic compounds as they are metabolized.[14]

Inhalation of irritant compounds in sufficient concentrations induces reflex changes in breath frequency and tidal volume, as well as changes in the shapes of inspiratory and expiratory flow vs. time curves.[13,18-24] These responses can be highly variable, depending upon the concentration and composition of the inhaled irritant agent. Pulmonary (deep lung) irritants, such as ozone, induce a rapid, shallow breathing pattern that is a vagally mediated reflex believed to originate by the stimulation of C fibers in the lung.[2,18,25-28] Compounds classified as sensory (or upper airway) irritants, such as volatile aldehydes, induce reflex responses mediated by trigeminal afferent pathways.[18,29] These reflexes generally involve decreases in frequency, accompanied by depression of ventilation or a compensatory increase in tidal volume to slow, deep breathing.[21,30] Reflex breathing pattern changes have been interpreted as pulmonary defenses against respiratory injury. In addition to the reduction in inhaled dose rate that results from depression of ventilation, reciprocal changes in frequency and tidal volume that do not change $\dot{V}_E$ will modify the distribution of the inhaled dose within the respiratory tract. Changes in dose distribution could be protective of particularly sensitive tissues. The rapid, shallow breathing pattern response to a relatively low solubility pulmonary irritant, such as ozone, could shift deposition away from more distal lung alveoli. The slow and deep breathing pattern response to highly soluble upper airway irritants, such as formaldehyde and sulfur dioxide, may increase absorption of these water-soluble compounds in the upper airways and thus protect deep lung tissues.

II. Problems Encountered

A. Restrained vs. Unrestrained Animals

A common goal of laboratory inhalation exposures is to duplicate natural human exposure conditions as closely as possible. It is therefore desirable to avoid physical restraint and the outfitting of subjects with encumbering apparatus during exposure; however, this can conflict with other important objectives. Some level of confinement is required to conduct an exposure experiment. As discussed in Chapter 5, animals may be placed in cages inside an exposure chamber to receive controlled concentrations of airborne pollutants. The difficulty with exposure systems in which animals are unrestrained is that they can maneuver to avoid exposure. They may move to regions of the cage that have low airflow delivery and thus lower concentrations of the pollutants, and the animals may additionally avoid exposure by tucking their heads into their fur or by sleeping (which decreases ventilation).

Physiological monitoring is a more complex technical problem when applied to unrestrained animals. Metabolic gas exchange can be readily measured if the input and output airflow to the subject is confined and sampled (see Section IV.C). Body temperature, heart rate, and other physiological parameters may be measured by radiotelemetry from surgically implanted sensors; however, this requires the use of invasive animal preparations in advance of exposure.[31] Minute ventilation and breathing pattern cannot be readily measured, although breathing movements can be counted visually or recorded using radiotelemetry. If the exposure chamber containing the animal is relatively small and can be nearly completely sealed, then small pressure pulses associated with breathing pattern can be measured (see Section II.C). If the objective is to measure the effect of exposure on spontaneous locomotion of an animal, then use of an exposure arena without any restraint is essential.

By using various forms of restraint, additional physiological monitoring is possible. For animals held in a fixed position, core body temperature can be measured with rectal temperature probes; however, restraint and the presence of the probe may affect temperature regulation in rodents due to their labile thermoregulatory physiology. By enclosing the body of an animal in a tube or small chamber and holding the animal at the neck by a soft collar seal, breathing pattern and ventilation can be easily measured (see Section IV).[24,32] For large animals, masks of various designs can be used to confine and measure the flow of respiratory air.[33-34] Restraining animals or outfitting them with masks may affect breath frequency and tidal volume and may or may not affect minute ventilation.[35] For any physiological measurements made during an exposure, matched purified air controls that receive a matched-duration exposure are necessary in order to distinguish between effects of the exposure pollutant and effects of restraint or attachment to the exposure apparatus that may develop over the time-course of the exposure.

B. Nose-Only Exposure

Effective control of inhalation dose-delivery in exposures to reactive pollutants requires a nose-only exposure. Reactive pollutants may be scrubbed from the exposure atmosphere by contact with animal fur, urine, or feces. Airborne acids are readily neutralized by ammonia evolving from urine and feces, and particles may be captured by electrostatic interactions with fur. As mentioned in Chapter 5, nose-only exposure, which usually includes both the mouth and nose, involves holding an animal in a tube so that a stream of air with the pollutant flows past the respiratory orifices. This preparation is easily augmented to permit measurement of breathing pattern and minute ventilation, and once the animal is restrained, access is provided for additional physiological measurements.

C. Measuring Tidal Volume

Tidal volume can be measured in several ways, each of which presents different problems when conducting inhalation exposures. The most straightforward technique used with small animals is flow plethysmography. In a flow plethysmograph, the animal's body is contained in a small tube or chamber, and the respiratory orifices are isolated from the thorax by a collar seal around the neck.[35] Breathing movements of the thorax move air into and out of the body chamber through a laminar flow resistance element, which can be a commercially available pneumotachograph or a series of fine-mesh screens in a tubular enclosure. Small pressure changes associated with the movement of air across the resistance element are measured with a sensitive, calibrated pressure transducer, and the pressure signal is integrated over each breath cycle to provide a measure of the tidal volume as the volume displacement of air around the thorax. The pressure changes are too small to significantly involve compressive heating of air, and any small leaks around the neck seal will not result in a large diversion of air flow from the resistance element if the thoracic containment and resistance element comprise a sufficiently low-pressure system during breathing.

Alternatively, in a pressure plethysmograph, the thoracic enclosure chamber is completely sealed with no low-resistance flow path, and thoracic expansion and contraction cause pressure changes in the chamber which are measured with a pressure transducer. The chamber must be large enough so that breathing will not be impeded by pressure changes. This preparation, however, is more sensitive to leaks, and rapid breathing may cause significant errors in the calibration of the pressure signal due to compressive heating of the air. Changes in heat transfer to the chamber, due to changes in animal body temperature, may also significantly affect the calibration.

The barometric plethysmograph accommodates an unrestrained animal. The entire animal is confined in a sealed chamber, and pressure changes in the chamber associated with heating and humidification of inhaled air are recorded. With additional measurements of body temperature, chamber humidity, and chamber temperature, tidal volume can be derived.[36] Despite the attraction of making the measurement using an unrestrained animal, this method is difficult to use accurately. Inspired air may be warmed to core body temperature, but expired air leaves the nasal passages saturated with water vapor at a temperature intermediate between chamber air and body temperature.[37-41] Room pressure changes affect the measurement, and closure of the chamber to make the measurement interrupts the exposure, initiates changes in respiratory gas fractions due to O_2 and CO_2 exchange, changes heat transfer from the animal to the chamber to the outside air, and may also disturb the animal, thereby affecting its breathing pattern.

Respiratory masks provide for measurement of respiratory flow rate and tidal volume by separating inspired and expired air with valves. They are most useful for large animals, such as dogs and horses, and for humans, since these species are easily trained to wear masks, have large tidal volumes in relation to the dead space volume of a mask, and have large respiratory flow rates which readily overcome the resistance of the mask valves.[33,42]

III. Scope and Limitations

A. Small and Large Animal Exposures

The choice of a smaller vs. a larger animal model for inhalation exposure studies offers different sets of problems and opportunities for exposure equipment design and physiological monitoring. Large animals will usually require large chambers or mask exposure systems; however, they can be outfitted with a wide variety of physiological instrumentation. Large subjects, including humans and dogs, are more intelligent then rodents and can be trained to wear physiological recording apparatus and to engage in complex behavior (such as exercise) during exposure.[33,43] In contrast, small animals can be exposed and monitored in larger numbers, and, due to their shorter life spans, chronic exposures can be performed in months, thus spanning a large fraction of their lifetimes. The monitoring techniques for large animals is beyond the scope of this chapter, and the focus here is on physiological monitoring of small animals during exposures.

B. Other Physiological Monitoring

The physiological variables addressed in this chapter for exposure monitoring are respiratory parameters that are related to characterizing inhaled dose rate, including breathing pattern, minute ventilation, and metabolic rate. There is a large variety of other physiological measurements that can be made when examining the effects of inhaled pollutants. These include measures of airway resistance and pulmonary compliance, reactivity of airways, shape and timing of tidal air flow curves, ventilatory responses to an elevated carbon dioxide concentration, lung volume changes, lung gas diffusion properties, maximum expiratory volume and flow characteristics, inert gas washout dynamics, and oscillatory mechanics of the respiratory system. Most pulmonary function tests developed for humans have been adapted, with various degrees of success, to small animals, and the reader is referred to the related literature for discussions of these techniques.[24,32,44]

IV. Respiratory Monitoring

A. Breathing Pattern

Breath frequency is an easily measured respiratory variable, and it is also among the most sensitive with respect to responses to inhaled irritant pollutants. Combined with a measurement of tidal volume, minute ventilation (a measure of inhaled dose rate) can be obtained. (The discussion below includes the measurement of tidal volume; if breath frequency alone is desired, breathing movements can be recorded simply with a pressure transducer attached to a tube containing the animal.) In a typical preparation, the animal subject is held in a plastic tube, and the head chamber is separated from the body by a membrane collar of latex dental dam (Figure 6.1). A cylindrical tube with a diameter slightly larger than that of the body of the animal is effective as a body chamber. It permits breathing movements, but restricts large body movements. The rear end of the tube is sealed with a rubber stopper, and a backstop plate on a sliding rod can be adjusted to accommodate the length of the body. An opening in the body chamber allows air displaced by thoracic breathing movements to move into and out of the chamber. This opening must be placed where it cannot be blocked by the animal's body. In the arrangement shown in Figure 6.1, the body tube is perforated around the thorax, and the openings are enclosed by an outer sleeve with a single opening and sealed to the body tube by O-rings. The flow rate of the displaced air is measured using a calibrated resistance element fitted with a pressure transducer. Shown in the figure is a commercial pneumotachograph (OEM Medical; Richmond, VA), although laminar flow resistance elements for rodents can be constructed with several layers of 500-mesh stainless steel screens in a cylindrical tube 12 mm in diameter. The resistance to flow should not be large enough to interfere with breathing, and, in practice, a flow resistance element with resistance of 0.02 cm sec^{-1} ml^{-1} is adequate for young adult rats (about 250 g in mass). A differential pressure transducer connected across the resistance element produces a signal proportional to the flow rate. The resistance element is connected between the body tube and the head chamber. By using the head chamber to provide the reference or baseline pressure (instead of room air, for example),

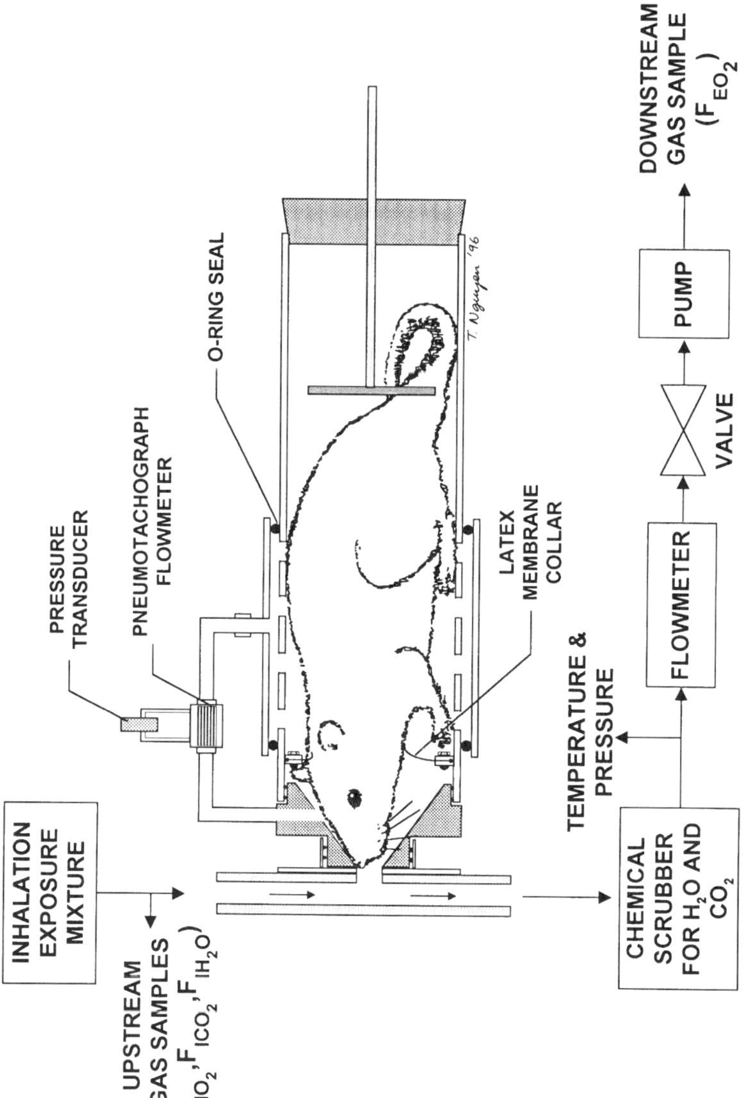

Figure 6.1
Breathing pattern and metabolic rate measurement system for rats during nose-only inhalation exposure.

one avoids addition of a baseline flow of air between the head and body chamber due to any small difference in average pressure between the exposure air stream and the room, and one also minimizes the effect of transient changes in room air pressure from air conditioning systems or the opening and closing of doors. The transducer (flow) signal is then electrically or digitally integrated to yield the volume of each breath, and the signal cycles are counted to measure breath frequency. Minute ventilation is obtained as the product of tidal volume and breath frequency.

Breathing pattern and minute ventilation may vary greatly during the course of an exposure. Body movements of the animal increase metabolic gas exchange and $\dot{V}_E$, and breathing at constant $\dot{V}_E$ may involve transient shifts from more rapid and shallow to more slow and deep breathing patterns. To accurately characterize breathing patterns during exposure, it is necessary to take repeated samples for averaging. A typical sampling protocol might include a 1-minute sample of steady-state breathing every 20 minutes over a 3-hour exposure period.[13]

At the beginning of an exposure, animals are likely to be excited as a result of handling and being loaded into exposure tubes. An initial recovery period of about 30 minutes allows them to settle down to more consistent breathing patterns. Nevertheless, breathing patterns may still change progressively over the course of a purified air exposure lasting an hour or more. Therefore, the practice of making an initial baseline control measurement from an animal breathing purified air and comparing a subsequent series of pollutant exposure measurements to this single initial measurement should be avoided. A matched set of purified air control animal exposures that extend through the full duration of the experimental pollutant exposures is the proper basis for comparison of breathing patterns.

B. Minute Ventilation and Inhaled Dose Rate

Tidal volume and minute ventilation are expressed as expired volume, or expired volumetric flow, of air under specified conditions of body temperature and pressure and with air saturated with water vapor (BTPS). The reason for this distinction is that the inhaled volume of air entering the respiratory system, at variable ambient temperatures and humidities, is heated and humidified to relatively constant conditions (of water vapor saturation and body temperature) in the lungs. At the maximum inspiratory volume and the maximum expiratory volume, the respiratory airflow rate is zero, and the air in the lungs is at a pressure equal to that of the outside air. The volume of air displaced into and out of the body chamber by movements of the thorax is equal to tidal volume at BTPS. The volume of air actually leaving the mouth and nose on expiration is slightly smaller due to cooling and condensation of some water vapor in the upper respiratory tract, but this small change is usually negligible. However, as a measure of inhaled dose rate of a pollutant, $\dot{V}_E$ BTPS may need to be corrected to estimate $\dot{V}_I$, the inspired flow rate of air containing the pollutant at ambient temperature and humidity.[13] For example, an animal with a body temperature of 37°C and an expired $\dot{V}_E$ of 200 m*l*/min BTPS, inhaling ambient air at 20°C and 30% relative humidity, inhales the ambient air at 179 m*l*/min, or about 10% less than the volume flow at BTPS. Unequal exchange of O_2 and CO_2 also contributes to the mismatch between $\dot{V}_E$ and $\dot{V}_I$, but the contribution is small and variation in ambient temperature and humidity has the larger effect under most conditions.

C. Respiratory Gas Exchange and Body Temperature

The respiratory gas exchange can be obtained by measuring the flow of air past an animal and the respiratory gas fractions in the air upstream and downstream of the animal. Oxygen consumption or carbon dioxide production, or both, may be measured, although $\dot{V}_{CO_2}$ is more variable than $\dot{V}_{O_2}$ because of transient changes in the pool of dissolved CO_2 in equilibrium with bicarbonate in body fluids.

Figure 6.1 shows a diagrammatic configuration for measuring $\dot{V}_{O_2}$ during an inhalation exposure. An oxygen analyzer measures the O_2 fraction in upstream and downstream air, and a flowmeter measures the flow rate of air downstream. It is essential when measuring oxygen consumption that entry or exit of all gases to and from the exposure gas stream be taken into account, as many effects often assumed to be negligible produce surprisingly large errors in the determination of gas exchange.[45] Thus, the flow of air downstream includes added water vapor evaporated from the animal and metabolic CO_2 produced by the animal, as well as decreased O_2 (consumed by the animal). There are a variety of methods to accomplish this accounting.[45-49] One of the most straightforward arrangements is shown in Figure 6.1. All flow rates are expressed at standard temperature (273 K) and pressure (1 atmosphere = 760 mmHg) and with water vapor-free (or dry) conditions (STPD). Air flow leaving the respirometer, $\dot{V}_{RE_{stpd}}$, is corrected to STPD conditions by the following:

$$\dot{V}_{RE_{stpd}} = \dot{V}_{RE_{atph}} \frac{(P_B - P_{H_2O})\,273}{760\,T_A} \tag{6.1}$$

where:

$\dot{V}_{RE_{atph}}$ = air flow out of the respirometer at ambient temperature, pressure, and humidity.
P_B = barometric pressure of the outflow airstream (mmHg).
P_{H_2O} = partial pressure of water vapor in the outflow airstream (mmHg).
T_A = ambient temperature of the outflow airstream (K).

Total pressure can be determined with a barometer, and relative humidity can be measured with a hygrometer, along with tables to convert to water vapor pressure.[50] Alternatively, water vapor can be scrubbed chemically from the respirometer outflow (also shown in Figure 6.1) using a desiccant so that $P_{H_2O} = 0$.

Oxygen consumption can then be determined using a measure of CO_2-free outflow air and the following equation:[46-48]

$$\dot{V}_{O_2} = \dot{V}'_E \frac{(F'_{IO_2} - F'_{EO_2})}{(1 - F'_{IO_2})} \tag{6.2}$$

where $\dot{V}'$ and F' are (respectively) CO_2-free air-flow rates and CO_2-free gas fractions of upstream (subscript I) and downstream (subscript E) air. The measurement of CO_2-free flow rate can be made by chemically scrubbing CO_2 from the outflow air before the flow rate measurement (Figure 6.1); CO_2-free oxygen fractions are measured by scrubbing CO_2 from the gas sample stream before it enters the O_2 analyzer. If water vapor is measured to mathematically correct the flow measurement (rather than scrubbing it), then water vapor should be scrubbed from the O_2 analyzer sampling lines before O_2 fractions are measured. Upstream gas fractions may be predicted from standard atmosphere values for dry air ($F_{CO_2} = 0.0003$, $F_{O_2} = 0.2095$, CO_2-free $F'_{O_2} = 0.2096$); however, respiratory gas fractions of air may be altered by chemical filtration to clean the air and by addition of pollutants into carrier gases for toxicology studies.[45] It is best to measure both upstream and downstream gas fractions.

It can be awkward to scrub H_2O and CO_2 from the exposure flow, particularly if the flow rates are large and several animals are being exposed in multiple respirometers. It is then easier to measure water vapor and F_{CO_2} and to correct flow measurements to yield H_2O- and CO_2-free flows. Once a CO_2 analyzer is introduced, then $\dot{V}_{CO_2}$ can be measured in addition to $\dot{V}_{O_2}$. Equations and configurations of an exposure apparatus for the combined determination of $\dot{V}_{O_2}$ and $\dot{V}_{CO_2}$ have been described.[45,49]

Body temperature can be measured during exposure with a rectal temperature probe for continuous monitoring or by using an implanted temperature telemetry system.[31] If continuous placement of a rectal probe or the invasive surgical preparatory procedure for telemetry is undesirable, body temperature can be measured simply and relatively noninvasively with brief rectal probe measurements made before and immediately following exposure.

V. Suppliers of Equipment

A. Rodent Plethysmograph Systems

Buxco Electronics, Inc.
635 Second Avenue
Troy, NY 12182

Columbus Instruments
950 North Hague Avenue
Columbus, OH 43204

B. Pressure Transducers and Pneumotachographs

Validyne Engineering Corp.
8626 Wilbur Ave.
Northridge, CA 91324

C. Flow, Temperature, and Relative Humidity Monitors

Fisher Scientific
711 Forbes Avenue
Pittsburgh, PA 15219

Physitemp
154 Huron Avenue
Clifton, NJ 07013

SKC West, Inc.
P.O. Box 4133
Fullerton, CA 92834-4133

Edgetech
455 Fortune Blvd.
Milford, MA 01757

D. Respiratory Gas Analyzers

Ametek
Thermox Instruments Division
150 Freeport Road
Pittsburgh, PA 15238

Columbus Instruments
950 North Hague Avenue
Columbus, OH 43204

E. Water and Carbon Dioxide Absorbants

W.A. Hammond Drierite Co.
138 Dayton Avenue
Xenia, OH 45385

Thomas Scientific
P.O. Box 99
Swedesboro, NJ 08085-6099

References

1. Morgan, M. S. and Frank, R., Uptake of pollutant gases by the respiratory system, in *Respiratory Defense Mechanisms*, Vol. 5, Part I, Brain, J. D., Proctor, D. F., and Reid, L. M., Eds., Marcel-Dekker, New York, 1977, chap. 6.
2. Adams, W. C., Savin, W. M., and Christo, A. E., Detection of ozone toxicity during continuous exercise via the effective dose concept, *J. Appl. Physiol.*, 51, 415, 1981.

3. Valberg, P. A., Brain, J. D., Sneddon, S. L., and LeMott, S. R., Breathing patterns influence aerosol deposition sites in excised dog lungs, *J. Appl. Physiol.,* 53, 824, 1982.
4. Bennett, W. D., Messina, M. S., and Smaldone, G. C., Effect of exercise on deposition and subsequent retention of inhaled particles, *J. Appl. Physiol.,* 59, 1046, 1985.
5. Mautz, W. J., McClure, T. R., Reischl, P., Phalen, R. F., and Crocker, T. T., Enhancement of ozone induced lung injury by exercise, *J. Toxicol. Environ. Health,* 16, 841, 1985a.
6. Mautz, W. J., Kleinman, M. T., Phalen, R. F., and Crocker, T. T., Effects of exercise exposure on toxic interactions between inhaled oxidant and aldehyde air pollutants, *J. Toxicol. Environ. Health,* 25, 165, 1988.
7. Gerrity, T. R., Weaver, R. A., Berntsen, J., House, D. E., and O'Neil, J. J., Extrathoracic and intrathoracic removal of O_3 in tidal-breathing humans, *J. Appl. Physiol.,* 65, 393, 1988.
8. Punte, C. L., Owens, J. E. J., Krackow, E. H., and Cooper, P. L., Influence of physical activity on the toxicity of aerosols and vapors, *A.M.A. Arch. Indust. Health,* 17, 34, 1958.
9. Gardner, D. E., Illing, J. W., Miller, F. U., and Coffin, D. L., Enhancement of effect of exposure to O_3 and NO_2 by exercise, *Toxicol. Appl. Pharmacol.,* 29, 129, 1974.
10. Fukase, O., Isomura, D., and Watanabe, H., Effects of exercise on mice exposed to ozone, *Arch. Environ. Health,* 33, 198, 1978.
11. Malek, D. E. and Alarie, Y., Ergometer within a whole-body plethysmograph to evaluate performance of guinea pigs under toxic atmospheres, *Toxicol. Appl. Pharmacol.,* 101, 340, 1989.
12. Zeltner, T. B., Sweeney, T. D., Skornik, W. A., Feldman, H. A., and Brain, J. D., Retention and clearance of 0.9 micron particles inhaled by hamsters during rest or exercise, *J. Appl. Physiol.,* 70, 1137, 1991.
13. Mautz, W. J. and Bufalino, C., Breathing pattern and metabolic rate responses of rats exposed to ozone, *Resp. Physiol.,* 76, 69, 1989.
14. Gordon, C. J., Mohler, F. S., Watkinson, W. P., and Rezvani, A. H., Temperature regulation in laboratory mammals following acute toxic insult, *Toxicology,* 53, 161, 1988.
15. Tepper J. S., Wiester, M. J., Weber, M. F., and Menache, M. G., Measurements of cardiopulmonary response in awake rats during acute exposure to near-ambient concentrations of ozone, *J. Appl. Toxicol.,* 10, 7, 1990.
16. Watkinson, W. P. and Gordon, C. J., Caveats regarding the use of the laboratory rat as a model for acute toxicological studies: modulation of the toxic response via physiological and behavioral mechanisms, *Toxicology,* 81, 15, 1993.
17. Bartlett, R. G., Bohr, V. C., and Helmendach, R. H., Comparative effect of restraint (emotional) hypothermia on common laboratory animals, *Physiol. Zool.* 29, 256, 1956.
18. Alarie, Y., Sensory irritation by airborne chemicals, *CRC Crit. Rev. Toxicol.,* 2, 299, 1973.

19. Widdicombe, J. G., Respiratory reflexes and defense, in *Respiratory Defense Mechanisms*, Vol. 5, Part II, Brain, J. D., Proctor, D. F., and Reid, L. M., Eds, Marcel-Dekker, 1977, chap. 16.
20. Coggins, C. R. E., Musy, C., and Ventrone, R., Changes in the minute ventilation of rats exposed to different concentrations of cigarette smoke, *Toxicol. Lett.*, 11, 181, 1982.
21. Chang, J. C. F., Steinhagen, W. H., and Barrow, C. S., Effect of single or repeated formaldehyde exposure on minute volume of B6C3F1 mice and F-344 rats, *Toxicol. Appl. Pharmacol.*, 61, 451, 1981.
22. Dallas, C., Theiss, J. C., Harrist, R. B., and Fairchild, E. J., Respiratory responses in the lower respiratory tract of Sprague-Dawley rats to formaldehyde inhalation, *J. Environ. Path. Toxicol. Oncol.*, 6, 1, 1986.
23. Schaper, M. and Detwiler, K., Evaluation of the acute respiratory effects of aerosolized machining fluids in mice, *Fund. Appl. Toxicol.*, 16, 309, 1991.
24. Vijayaraghavan, R., Schaper, M., Thompson, R., Stock, M. F., and Alarie, Y., Characteristic modifications of the breathing pattern of mice to evaluate the effects of airborne chemicals on the respiratory tract, *Arch. Toxicol.*, 67, 478, 1993.
25. Coleridge, J. C. G., and Coleridge, H. M., Afferent vagal C fibre innervation of the lungs and airways and its functional significance, *Rev. Physiol. Biochem. Pharmacol.*, 99, 1, 1984.
26. McDonnell, W. F., Horstman, D. H., Hazucha, M. J., Seal, E., Haak, E. D., Salaam, S. A., and House, D. E., Pulmonary effects of ozone exposure during exercise: dose-response characteristics, *J. Appl. Physiol.*, 54, 1345, 1983.
27. Beckett, W. S., McDonnell, W. F., Horstman, D. H., and House, D. E., Role of the para-sympathetic nervous system in acute lung response to ozone, *J. Appl. Physiol.*, 59, 1879, 1985.
28. Schelegle, E. S., Carl, M. L., Coleridge, H. M., Coleridge, J. C. G., and Green J. F., Contribution of vagal afferents to respiratory reflexes evoked by acute inhalation of ozone in dogs, J. Appl. Physiol., 74, 2338, 1993.
29. Ulrich, C. E., Haddock, M. P., and Alarie, Y., Airborne chemical irritants: role of the trigeminal nerve, *Arch. Environ. Health*, 24, 37, 1972.
30. Davis, T. R. A., Battista, S. P., and Kensler, C.J., Mechanism of respiratory effects during exposure of guinea pigs to irritants, *Arch. Environ. Health*, 15, 412, 1967.
31. Watkinson, W. P., Wiester, M. J., and Highfill, J. W., Ozone toxicity in the rat. I. Effect of changes in ambient temperature on extrapulmonary physiological parameters, *J. Appl. Physiol.*, 78, 1108, 1995.
32. Costa, D. L. and Tepper, J. S., Approaches to lung function assessment in small mammals, in *Toxicology of the Lung*, Gardner, D. E., Crapo, J. D., and Massaro, E. J., Eds., Raven Press, New York, 1988, 147.
33. Stavert, D. M., Reischl, P., and O'Laughlin, B. J., A respiratory mask for resting and exercising dogs, *J. Appl. Physiol.*, 52, 500, 1982.
34. Gardner, D. E. and Kennedy, G. L., Methodologies and technology for animal inhalation studies, in *Toxicology of the Lung*, Second ed., Gardner, D. E., Crapo, J. D., and McClellan, R. O., Eds., Raven Press, New York, 1993.

35. Mauderly, J. L., Respiration of F344 rats in nose-only inhalation exposure tubes, *J. Appl. Toxicol.*, 6, 25, 1986.
36. Drorbaugh, J. E. and Fenn, W. O., A barometric method for measuring ventilation in newborn infants, *Pediatrics*, 16, 81, 1955.
37. Schmidt-Nielsen, K., Hainsworth, F. R., and Murrish, D. E., Counter-current heat exchange in the respiratory passages: effect on water and heat balance, *Resp. Physiol.*, 9, 263, 1970.
38. Jacky, J. P., A plethysmograph for long-term measurements of ventilation in unrestrained animals, *J. Appl. Physiol.*, 45, 644, 1978.
39. Jacky, J. P., Barometric measurement of tidal volume: effects of pattern and nasal temperature, *J. Appl. Physiol.*, 49, 319, 1980.
40. Epstein, M. A. F. and Epstein, R. A., A theoretical analysis of the barometric method for measurement of tidal volume, *Respir. Physiol.*, 32, 105, 1978.
41. Epstein, R. A., Epstein, M. A. F., Haddad, G. G., and Mellins, R. B., Practical implementation of the barometric method for measurement of tidal volume, *J. Appl. Physiol.*, 49, 1107, 1980.
42. Mauderly, J. L., Evaluation of the grade pony as a pulmonary function model, *Am. J. Vet. Res.*, 35, 1025, 1974.
43. Stavert, D. M., Reischl, P., and Murdock, L. C., A refrigerated treadmill apparatus for exercising dogs, *J. Appl. Physiol.*, 52, 495, 1982.
44. Costa, D. L., Interpretation of new techniques used in the determination of pulmonary function in rodents, *Fund. Appl. Toxicol.*, 5, 423, 1985.
45. Mautz, W. J., Calibration of respiratory gas exchange measurements in inhalation toxicology studies, *Fund. Appl. Toxicol.*, 18, 144, 1992.
46. Depocas, F. and Hart, J. S., Use of the Pauling oxygen analyzer for measurement of oxygen consumption of animals in open-circuit systems and in a short-lag, closed-circuit apparatus, *J. Appl. Physiol.*, 10, 388, 1957.
47. Hill, R. W., Determination of oxygen consumption by use of the paramagnetic oxygen analyzer, *J. Appl. Physiol.*, 33, 261, 1972
48. Withers, P. C., Measurement of $\dot{V}_{O_2}$, $\dot{V}_{CO_2}$, and evaporative water loss with a flow-through mask, *J. Appl. Physiol.*, 42, 120, 1977.
49. Mautz, W. J., Phalen, R. F., McClure, T. R., and Bufalino, C., A rodent treadmill for inhalation toxicologic studies and respirometry, *J. Appl. Physiol.*, 58, 673, 1985.
50. List, R. J., *Smithsonian Meteorological Tables*, Smithsonian Institution Press, Washington D.C., 1949.

Chapter 7

Tissue Acquisition and Processing

Ronald Rasmussen

Contents

0-8493-3343-1/97/$0.00+$.50

I. Introduction

A. Scope and Limitations

Because the preparation of biological tissue specimens for microscopic examination remains something of an art, it is not really possible to prepare a "recipe" which can be followed with the expectation of 100% success on the first attempt. Therefore, this chapter will provide only a general introduction to the basic procedures, focusing on the lung as the tissue of interest. Because of space limitations, a complete listing of all possible fixation, sectioning, and staining procedures also is not possible, and the investigator is directed to the appropriate references at the end of this chapter. To become proficient in tissue preparation, there is no substitute for practice, and, if possible the guidance of an experienced histologist.

B. Tissue Fixation

In order to perform extensive analyses of respiratory tract tissues, it is necessary to treat them in a way that preserves their gross and microscopic features.

This usually involves treatment with chemicals that denature or precipitate proteins *in situ* so that the interrelationships of tissue structures are maintained during subsequent preparation for microscopic viewing. The most popular fixatives traditionally have been aldehydes and alcohols, or their mixtures, which in combination denature protein and dehydrate the tissue. The choice of a particular fixative depends on the nature of the subsequent analyses.

1. Fixation for light microscopy

Buffered (with respect to pH) solutions of formaldehyde or glutaraldehyde, described below, are most frequently used as fixatives when tissues are to be examined only by light microscopy. The fixation is rapid, and tissues can be processed within a few hours. If the goal is to examine tissues for gross changes or pathology, these methods are adequate. However, rapid fixation may introduce artifacts that become apparent when the tissue is examined using quantitative measures or by electron microscopy.

2. Fixation for electron microscopy

The goal here is to preserve not only the gross morphology of tissue, but also the subcellular structures and their correct anatomical relationships. Thus, fixatives for electron microscopy (EM) must be rapid but also gentle as far as the perturbation of tissue structures is concerned. This can be accomplished by careful attention to the chemical purity, pH, and osmolarity of the fixative solutions. EM fixation often includes treatment with electron-dense metals such as osmium, tungsten, or gold, which bind to tissue membranes and render them visible when using an electron microscope.

C. Tissue Sectioning and Staining

1. Light microscopy

For routine work, tissues are embedded in specially formulated paraffin, and sections are cut using specifically designed equipment. For rapid processing, tissue can be quick-frozen and sectioned, again with the proper equipment. Sections are mounted on glass slides, the embedding medium is removed, and the tissue is stained for viewing.

Except for pigmented cells, tissue membranes and subcellular structures are nearly colorless. To make them visible in the light microscope, tissue sections can be treated with a wide variety of synthetic and natural pigments which react chemically with cellular structures and render them visible. While unstained tissue sections can be examined by phase contrast microscopy, the judicious use of stains having known chemical properties can elucidate the chemical nature of subcellular structures. Examples of frequently used stains are given below.

2. Electron microscopy

To take advantage of the resolving power of the electron microscope, tissue sections are cut as thinly as possible, usually to a fraction of a micrometer. This requires embedding them in a rigid medium so that the tissue will not be distorted during sectioning. Several formulations of plastic are available which meet this requirement, and references to examples are given below. The sections are then mounted so that the electron beam can pass directly through the specimen without traversing any intervening material, such as glass. Because tissue structures do not differ greatly in their absorption of electrons, it is customary to "stain" the tissue sections with metals which strongly absorb electrons. Some metals, such as osmium, have an affinity for cellular membranes and bind strongly to them. When the electron beam passes through the osmium-stained specimen, the membranes will absorb electrons more strongly and, in a photographic negative, appear light against a dark background and the reverse in a positive print. In this way, the tissue structures are made more visible.

D. Morphometric Analysis of the Lung

This type of analysis includes direct measurement of lung structures and cellular components with the goal of detecting more subtle effects of inhaled toxicants than are apparent by qualitative analysis. Preparation of the tissue for morphometry requires careful attention to fixation, sectioning, and staining so that distortion is minimized. An in-depth discussion of the related methods is outside the scope of this chapter, but several excellent reviews have been published.[1-3]

II. Common Problems and Pitfalls

A. Laboratory Safety

Various chemical and physical hazards are associated with the preparation of tissues for analysis. Some of the more important of these are discussed below; however, the best means for prevention of injury is accomplished by a rigorous safety training program for all persons using the laboratory facilities on either a temporary or long-term basis. At a minimum, this program should include a review of chemical and equipment storage, first aid equipment, fire control equipment, emergency communications, and procedures for use of any automated equipment.

1. Chemical hazards

Formaldehyde (HCHO) and glutaraldehyde (glutaric dialdehyde) are volatile and highly irritating to the eyes and mucous membranes at air concentrations greater than about 1 part per million (ppm). Repeated exposure to HCHO may result in sensitization to its vapor and may also carry a small risk of nasal cancer. Some fixatives contain cacodylic acid (dimethylarsinic acid) which is highly poisonous and may be a carcinogen. Osmium tetroxide (osmic acid) is frequently used in fixation and staining of tissues for electron microscopy and is also highly poisonous and volatile. Therefore, preparation and use of solutions must be done in a certified chemical fume hood equipped with a reliable indicator (pressure gauge, etc.) of the proper functioning of the hood. If occasional exposure to the vapors cannot be avoided, an appropriate respirator and impermeable plastic goggles should be used. Direct skin contact also should be avoided by the use of impervious gloves and laboratory aprons. Other chemicals that are occasionally used in fixative formulas are methanol and ethanol, both highly flammable and toxic. The dehydration of tissue in preparation for embedding or staining requires the use of highly flammable and toxic solvents including acetone, alcohols, toluene, and xylene. Therefore, a histology laboratory must provide adequate ventilation and fume hoods for fixation, embedding, and staining operations.

Picric acid (2,4,6-trinitrophenol), used to prepare Bouin's fixative and in some stains, is both highly corrosive and explosive when allowed to dry; it should be stored under water. To weigh quantities of the acid, small amounts are placed on filter paper to allow most of the water to drain off, and the material is weighed while moist. The rim of the container should be kept clean of any crystals, since an explosion hazard may exist if dry crystals are scraped or compressed by the lid.

2. Physical hazards

In any laboratory, broken glassware is a common hazard. Strong, properly fitting plastic goggles should be worn at all times when working with glass in a manner such that breakage hazards could exist. In the histology laboratory, a special hazard is the extremely sharp microtome knife which is exposed during use. These knives, whether disposable or reusable, should not be left mounted in the microtome when it is not in use.

Some infiltration and embedding procedures involve the use of vacuum ovens or chambers. If glass desiccators are used, they should be enclosed in protective shields in case of implosion. Electrically powered automatic infiltration and embedding apparatus should be protected from liquid spills (chemicals or water) in order to avoid the risk of electrical shock.

B. Tissue Autolysis Prior to Fixation

When a tissue specimen is removed from a living animal, or an animal that has expired, the process of autolysis (tissue breakdown) begins immediately. This is due to proteolytic enzymes and nucleases present in the tissue itself and also to unavoidable microbial contamination. Autolysis can be slowed by refrigeration, but if tissue sections are to be prepared from either quick-frozen or fixed tissue, the processing should begin as soon as possible. This is especially important in preparing tissues for electron microscopy, since the autolytic process quickly disrupts subcellular structures such as membranes, organelles, and nucleic acids.

C. Incomplete Fixation

Proper fixation of tissue requires rapid and complete penetration of the specimen by the fixative. Since penetration of the specimen depends largely on diffusion and is hence time dependent, penetration must be completed before significant autolysis occurs. Therefore, specimens should be small (no more than 0.5 cm in the smallest dimension) or should be fixed in a way that ensures rapid penetration of the tissue, consistent with maintenance of morphology. For example, rodent (or small animal) lungs are often fixed by intratracheal instillation or vascular perfusion of the fixative. These methods provide rapid contact of the fixative with all parts of the organ.

D. Overfixation

When tissues are left indefinitely in fixatives containing formaldehyde, they tend to become harder with time. For some tissues, such as the lung, this is not a great problem. However, for others, such as liver specimens, the tissue may become so hard that sectioning is very difficult. Therefore, when the specimens have been thoroughly penetrated by the fixative or if tissues are to be stored for more than a few days before embedding and sectioning, they should be transferred to a neutral-buffered saline solution.

E. Chemical Modification of Tissue by Fixation

Formaldehyde and glutaraldehyde act by chemically crosslinking and precipitating proteins *in situ*. Fixatives containing metals such as mercury (e.g., Zenker's) also act by protein precipitation. The selection of a particular fixative depends on the nature of the features of the specimen to be examined. It should be recognized, however, that fixation by any method changes the chemical nature of the tissue and can greatly affect the behavior of the tissue in subsequent procedures such as staining or immunological reactivity. For example, fixation with glutaraldehyde will result in unreacted aldehyde groups in the tissue. If it is desired to stain for polysaccharides using the periodic acid-Schiff (PAS) method, it is necessary to first reduce the aldehyde groups with a strong reducing agent such as sodium borohydride (see below). If immunocytochemical detection of antigenic markers is to be done, then the fixation should be the minimum necessary to stabilize the tissue. For cell cultures on glass slides, this will involve only a few minutes of fixation with 2% formaldehyde. Longer fixation will tend to destroy enzymatic activity and the original antigenic character of proteins or other biomolecules. However, once tissue has been embedded in paraffin or plastic, most biomolecules are relatively stable.

F. Poor Infiltration of Tissue by Embedding Media

Adequate infiltration of the embedding medium into the specimen is a prerequisite to obtaining suitable tissue sections with a microtome. To accomplish this, the tissue specimen should be no more than 0.5 cm thick in the smallest dimension, and sequential exposure to each of the components of the embedding media must be long enough to permit complete replacement of the previous component. Failure to observe this requirement can result in mushy, unsectionable tissue in the center of the specimen because of incomplete removal of water. As with fixation, overly long exposure to organic solvents will tend to harden tissue, especially dense tissue such as liver and kidney.

III. Necessary Basic Equipment

A. Chemical Fume Hood

A fume hood is indispensable when conducting procedures with potentially toxic solvents and gases. The hood should be certified to provide an air velocity at the working face of at least 0.5 m sec^{-1}. The hood should not be used as a storage cabinet; a separate, fireproof, ventilated cabinet should be used for storage of flammable solvents.

B. Fixation Apparatus

For most small specimens, simple immersion of the tissue in the fixative of choice is adequate. However, for proper fixation of lung tissue, an apparatus capable of maintaining the lung in the fully inflated condition during fixation is desirable. This ensures adequate uniform contact between the fixative and all parts of the lung and makes subsequent histologic and quantitative morphometric analyses possible. A description of such an apparatus capable of fixing up to 48 rodent lungs simultaneously has been published (described later in this chapter).[10] However, for fewer lungs, individual setups may be used. Such an arrangement is described below. In any case, experimental and control lungs should be processed simultaneously.

C. Tissue Processor

Programmable tissue processors are available from several suppliers, and this equipment is highly desirable if large numbers of samples are to be processed. The dehydration and embedding solvents are automatically changed according to an adjustable program. To ensure adequate dehydration and infiltration, the specimens are placed into a chamber from which the air is evacuated during sequential exposure to the solvents and, ultimately, to specially formulated paraffin wax. Once the tissues are thoroughly infiltrated, the tissues are embedded for sectioning. Again, experimental and control specimens should be processed together.

D. Microtomes

The design of the microtome has changed little since the early years of this century.[4] An embedded or frozen specimen is held rigidly in a vise-like device and moved across an extremely sharp knife edge. Ideally, very thin and uniform sections are cut from the specimen. In practice, the thickness of the sections is determined by the nature of the embedding material and the sharpness of the knife edge. For paraffin-embedded materials, the lower limit of section thickness is 2 to 4 μm, but usually sections are cut at 5 to 6 μm. For transmission electron microscopy, which requires sections less than 1 μm in thickness, specimens must be embedded in a hard plastic, held in a specially designed precision microtome, and cut using a glass or diamond knife edge, which can be much sharper than a steel knife. The choice of microtome depends on the goals of the study.

E. Slide Warmer

This is an electrically heated platform used for drying paraffin sections that have been placed on microscope slides. It provides gentle warming at 35 to 45°C to remove all traces of water before dewaxing the specimens with organic solvents.

F. Microscope

A laboratory microscope should be available in the histology lab to permit examination of slides during some staining procedures. An inexpensive, student-grade compound microscope is generally adequate.

IV. Specific Techniques

A. Euthanasia Methods

1. Injectable anesthetic agents

The preferred method of euthanasia in studies of the rat respiratory system is intraperitoneal (i.p.) or intravenous (i.v.) injection of an overdose quantity of a veterinarian-approved barbiturate anesthetic solution. Several are commercially available, and they usually contain pentobarbital in combination with solubilizing agents and preservatives. They are relatively inexpensive and have the advantage of rapid action with minimal discomfort to the animals. Also, the tissues of the respiratory system are not significantly affected. The injections should be performed by trained personnel. Since barbiturates are controlled substances, they must be locked up, and complete records must be kept of use and inventory. Expired anesthetics must be promptly discarded, for both practical and ethical reasons.

To provide deep anesthesia in the rat, an injection of a commercial preparation of pentobarbital is given i.p. in the lower region of the abdominal cavity, avoiding the liver. The minimum useful anesthetic dose is reported as 50 mg kg^{-1}, and the LD_{50} is reported as 75 mg kg^{-1}.[5]

A complete review of available anesthetics and methods of euthanasia is beyond the scope of this chapter, but such information is available in a review by the American Veterinary Medical Association.[6] This publication reviews acceptable methods of euthanasia, as well as those that are considered unacceptable. Additional information is also included in Chapter 8 in this book.

2. Inhalation anesthetics

Although several inhalation anesthetics are available and are useful as anesthetics for nonterminal procedures, their use for euthanasia is not recommended if the respiratory tract is to be studied. Ethyl ether, although inexpensive, is used infrequently because of its explosion hazard, and it may cause changes in the lung parenchyma as a result of its irritant effects. Others, such as halothane, methoxyflurane, and enflurane, are expensive and may also cause lung irritation. Again, see Chapter 8 for a description of a single-animal unit for delivering methoxyflurane.

B. Lung Fixation by Intratracheal Perfusion for Light Microscopy

1. Fixative formulas

The choice of a fixative formula depends on the purposes of the study. In general, formaldehyde penetrates tissue rapidly, but produces complete fixation relatively slowly. Because of its bifunctional nature, glutaraldehyde fixes tissue rapidly, but penetrates tissue slowly. The pH and the osmolarity of the fixative solution will influence the ultimate micromorphology of the tissue, which can be very important in studies of subtle effects of airborne materials in the lung. Based on these principles, various formulations have been developed to combine the properties of formaldehyde and glutaraldehyde. The most widely used is probably "Karnovsky's Fixative". The original formulation was a hypertonic mixture (550 mOsm), but the designation is sometimes applied to any fixatives containing a mixture of formaldehyde and glutaraldehyde in various proportions in phosphate or cacodylate buffer of various osmolarities. A typical formula follows:[7]

Karnovsky's Fixative

Paraformaldehyde	40 g/*l*
Glutaraldehyde	50 g/*l*
$CaCl_2$	0.5 g/*l*
Cacodylic acid	12.8 g/*l*
Adjust to pH 7.2 with HCl	
Osmolarity: Approx. 550 mOsm	

To prepare the above stock fixative, the paraformaldehyde is suspended in a volume of deionized or distilled water of about 80% of the final volume at a temperature of 60°C. While stirring continously (e.g., with a magnetic stirring bar), 6 *N* NaOH is added dropwise over a period of several minutes, until the paraformaldehyde is completely dissolved. The total volume of NaOH added should be approximately 1 m*l* per 10 g of paraformaldehyde. After cooling at room temperature, the remaining ingredients are added; the pH is adjusted, if necessary, with HCl or NaOH; and the fixative is brought to the final volume with deionized water. The fixative is stable for at least several months when stored in the dark and refrigerated. To prepare the working fixative, the stock fixative is diluted 1:4.5 with 0.23 *M* cacodylic acid, and the pH is adjusted to 7.2 with HCl. This formulation is adequate for most light microscopic analyses and even for most transmission electron microscopic work.

2. Apparatus

To adequately fix lungs by intratracheal perfusion using the above fixative, three factors must be critically controlled. They are (1) rate of filling the lungs, (2) hydrostatic pressure during and subsequent to filling, and (3) time period that the pressure must be maintained.[8]

Because glutaraldehyde fixes tissue very rapidly, the lungs must be filled within a few seconds in order to fix them in the completely expanded state and maintain them in this state until fixation is complete. This can be accomplished for individual rat lungs by use of a disposable 50-m*l* syringe barrel with the plunger removed and replaced by a cork which has a narrow-bore glass or plastic tube through it, extending almost to the outlet fitting end of the syringe barrel. A rubber tube with a pinch clamp is attached to the syringe fitting, and fitted at the end with a male connector. The syringe barrel is filled with fixative, the cork is inserted, and the vertical distance between the end of the internal glass tube and the end of the male fitting attached to the rubber tube can then be set at the desired hydrostatic pressure.[9]

If several lungs are to be fixed simultaneously, the syringe can be replaced by a large reservoir to which is attached an arrangement of tubing; however, such an apparatus is adequate for only a few rats, since the flow rate of fixative through individual tubes will decrease as more tubes are added. An apparatus which overcomes this difficulty and can accommodate 48 rodent lungs at one time has been described by McClure et al.[10] This has a large fixative reservoir and a recirculation pump that can hold a constant pressure for several days (Figure 7.1).

3. Procedure for a single animal

The rat is deeply anesthetized, a tracheostomy is performed, and a 14-gauge stub cannula with a female fitting is inserted into the trachea and tied in place. The abdominal cavity is opened and the hemidiaphragms punctured or cut to collapse the lungs (take care not to damage the lungs themselves). The tube from the fixative reservoir is immediately attached to the stub cannula and the pinch clamp removed to allow fixative to rapidly fill the lungs (Figure 7.2). At a pressure of 20 to 30 cm of water, the lungs should inflate to approximately 80 to 90% of total lung capacity.[8] (At lower pressures, lung inflation is likely to be incomplete and the lungs will not be suitable for accurate morphometric analysis.) This pressure should be maintained for a minimum of 2 hours with glutaraldehyde-containing fixatives and for at least 2 days with formaldehyde-based fixatives. Adequacy of fixation can be judged by removing the reservoir tube from the tracheal cannula. If the lungs are fully fixed, little or no fixative should flow from the lungs out of the cannula.

C. Lung Fixation by Vascular Perfusion

Vascular perfusion of the lungs with fixative may be desirable to preserve the liquid lung-lining material and the locations of cells (e.g., macrophages and neutrophils) on the surfaces of the airways. Vascular fixation has the disadvantages of requiring more technical skill than intratracheal perfusion and requiring more time for adequate fixation of lung tissues.[11]

1. Fixative Formulas

The same fixatives are used as for intratracheal perfusion.

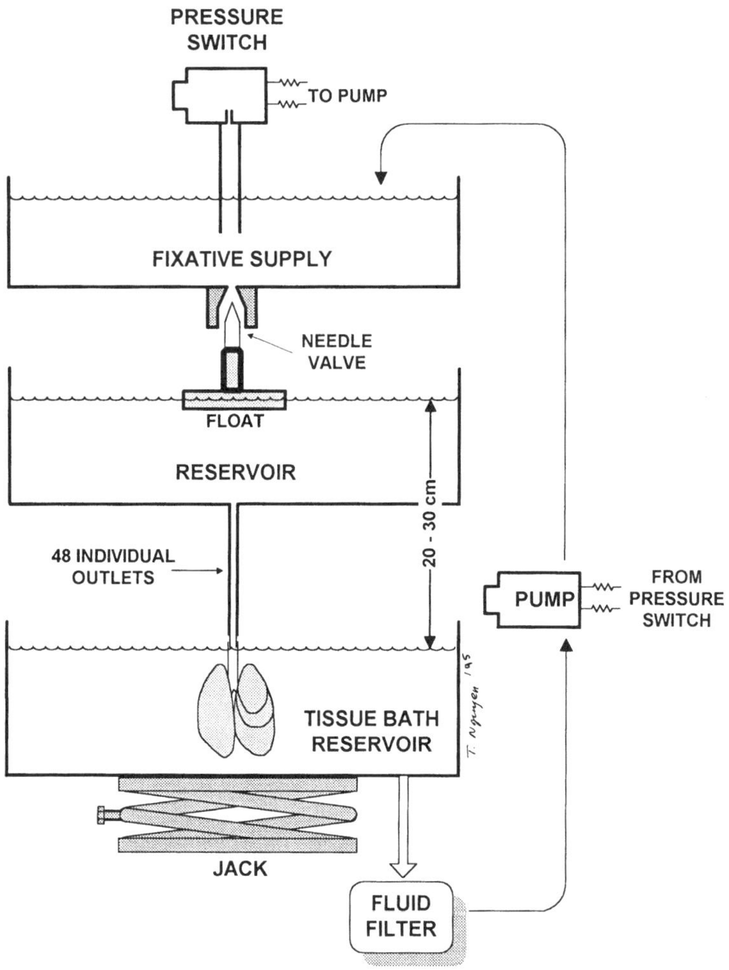

Figure 7.1
An airway perfusion apparatus for simultaneous fixation of up to 48 lungs. Details for construction and use are found in McClure, Diller, and Phalen.[10]

Equipment and Supplies

Physiological saline (0.9% NaCl) containing sodium heparin at 50 units/m*l* and prewarmed to 37°C
Fixative solution
Reservoirs for saline and fixative
Pinch clamp
3-way valve
Soft plastic tubing, 3–4 mm external diameter

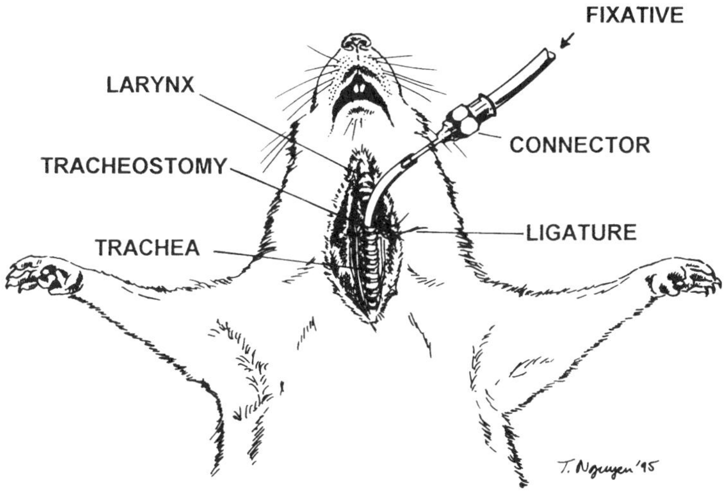

Figure 7.2
Rat with tracheostomy prepared for fixative infusion *in situ*.

14- or 15-gauge cannula with 2 cm of plastic tubing extension
20- or 21-gauge, 1" hypodermic needles
Male adapters to connect tubing to needles and cannula 10-m*l* syringe
Pentobarbital anesthetic

2. Procedure

This operation is best performed by two persons. Figure 7.3 is a diagram of the setup. The aim is to remove the blood from the vasculature and follow it immediately with a fixative solution. After deep anesthesia of the rat, the chest cavity is opened and the connector stub cannula with the tubing extension is placed in a tracheostomy and tied in place. At this point the heart should still be beating. The descending vena cava is cut to allow drainage of blood from the venous system. The 21-gauge needle attached to the tube from the solution reservoirs is inserted into the right ventricle of the heart and held in place. The pinch clamp is opened to permit saline to perfuse the vasculature. The lungs are deflated and inflated slowly 3 to 4 times with air via the tracheal cannula to a pressure of 20 to 25 cm of water and finally are held at this pressure. When the blood has been removed from the lungs, as indicated by their becoming almost completely white, perfusion of the vasculature with fixative is begun by switching the valve from the reservoirs and is continued at a pressure of 25 to 30 cm of water until the lungs are completely perfused with the fixative. With the lungs inflated with air, the trachea is then tied shut and the lungs removed to a container with fresh fixative to complete the process.

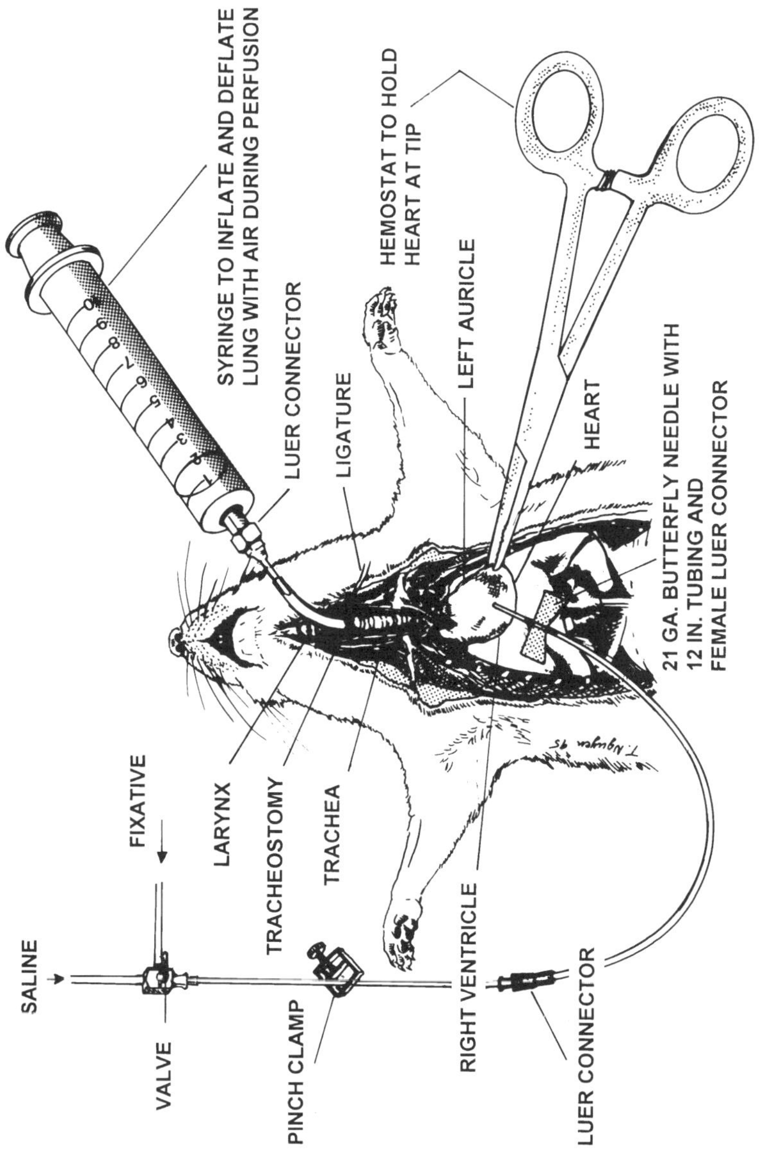

Figure 7.3
Rat prepared for vascular perfusion with fixative. Inflation and deflation of lungs with air facilitates even distribution of fixative in the lung vasculature.

D. Fixation of the Nasal Region

Depending upon the intended analyses, the fixative can simply be buffered 10% formaldehyde or, for scanning or transmission EM, Karnovsky's fixative, as described above. An important point is that the nasal cavity must be filled with the fixative to ensure contact with the epithelial cells covering the turbinates and olfactory region. This is usually accomplished by removing the animal's head and flushing fixative through the posterior opening of the nasopharynx and out through the nares. Some investigators immerse the head in fixative and briefly place it under vacuum to aid in the removal of air bubbles. When all air has been removed, the head is immersed in fixative for the required period.

Before embedding in paraffin or plastic and sectioning the nasal region, the bone must be decalcified. This can be accomplished using acids (such as formic acid); however, some loss of protein and nucleic acids will occur. To avoid this problem, many laboratories use a solution of 6% tetrasodium EDTA (edetate sodium; Versene) in 0.05 *M* Tris buffer at pH 7.6. After skinning and trimming them free of external muscle, the nasal sections of the rat heads are suspended in an amount of this solution which is at least 10 times the total volume of tissue. The solution is gently stirred at room temperature on a magnetic stirring plate and is changed weekly for 3 to 4 weeks, or until an examination indicates that the heads are decalcified sufficiently to permit sectioning. This can be confirmed by testing whether a sharp pin can easily penetrate the rat's incisor, or by determining whether a noncritical region of the skull can be easily cut through with a razor blade. The heads can then be grossly sectioned to isolate the regions of interest and embedded for microtome sectioning.

E. Fixation of Lung for Electron Microscopy

The fixative formulas are similar to those used for light microscopy; however, special attention must be paid to the purity and freshness of the chemicals. Commercial formalin (37 to 40% formaldehyde) is unsuitable, since it usually contains about 10% methanol as a preservative. Therefore, freshly prepared solutions of reagent grade paraformaldehyde must be used. After prolonged storage, solutions of both formaldehyde and glutaraldehyde tend to polymerize and become unusable. Specially purified reagents for fixation of tissue for electron microscopy are available from several chemical suppliers. Fixative solutions should be prepared on the day of intended use.

F. Fixation of Lung by Microwave Heating

1. Oven calibration and standardization

Fixation by microwave heating has been proposed as a rapid and efficient method of preparing tissue for microscopic analyses. However, experience has shown that in order to use this method, careful attention must be given to control and uniformity of the heating, as well as to the time of exposure of the specimen to microwave radiation. The procedure should not be considered as "cooking" the specimen, but rather as the application of just sufficient heating to stabilize protein-based structures in the tissue. To accomplish this goal, the characteristics of the particular oven being used must be determined with respect to distribution of microwave energy within the oven, and a region defined where a known energy density can be reproducibly delivered to the specimen being processed.[12] The size of the specimen to be fixed by this method is limited. Turner et al.[13] described a method for microwave fixation of whole rat lungs and indicated that artifacts associated with intratracheal instillation of fixatives could be avoided. In tests in our laboratory with whole rat lungs, the degree of preservation of tissue was not uniform throughout the entire lung. In some regions the fixation was quite good, while in others the tissue was essentially "cooked" and unsuitable for microscopic analysis. An extensive review of microwave fixation has been published recently,[12] and the reader is referred to this review for a detailed description of the methods.

G. Embedding, Sectioning, and Staining of Lung Tissue

1. Embedding and sectioning with paraffin

To prepare the fixed rat lung for microscopic analysis, it has been recommended that longitudinal (parallel to major bronchi) samples be taken from each major lobe which include a substantial portion of the main bronchus.[7,9] The samples should be no more than 3 to 4 mm thick to ensure adequate infiltration of the embedding medium. If it is desired to relate subsequent morphometric measurements to the original dimensions of the lung, then the fixed lung volume should be determined prior to sampling. This is done simply by measuring the weight of water displaced by the inflated lung.[14] The trachea is clamped, and the inflated lung is immersed in a beaker of water or buffer on a scale. If the density of the water is assumed to be 1 g ml^{-1}, the difference in weight with and without the lung present is the lung volume. During the embedding process, the tissue will shrink to a significant extent, in part due to loss of water. To determine the extent of shrinkage, the area of the surface to be sectioned is measured before embedding and after the first section has been cut and mounted. These measurements can be done most conveniently using an image analysis system. Alternatively, photographs can be taken of the cut face of the specimen before and after embedding, using the same magnification factor in each case. If morphometry is not to be done, then these preliminary measurements are unnecessary.

Although the dehydration and paraffin embedding can be done manually, it is strongly recommended that a tissue processor be used or that the embedding be done by a commercial histology laboratory. A typical embedding schedule for rat lung slices using an automatic processor is as follows. The slices are contained in tissue processing cassettes of a size such that the specimens are not compressed. They are then immersed sequentially with stirring for 90 min in each of the following percent concentrations of 2-propanol in water: 50, 70, 80, 95, 95, 100, 100; then two changes of xylene; and finally two changes of histological-grade paraffin (melting point, 55 to 57°C). From the final paraffin bath, the specimens are embedded with additional clean paraffin in the appropriate sized histology molds for sectioning. The temperature of the embedding paraffin should not exceed 65°C.

Because the lung specimens embedded in paraffin blocks may be up to 3 cm or more in length, they should be thoroughly chilled (to –20°C) before sectioning is attempted. Sectioning is done across the narrow dimension of the specimen. The blocks are mounted in a microtome and paraffin shaved off until full sections are obtained. At this point, the block is chilled further using "Histo-Freeze"™ * or its equivalent, and a series of 5-μm thick sections is cut as quickly as possible and floated on a 40 to 42°C distilled water bath containing a trace of gelatin (0.1 g l^{-1}). The sections are then picked up on clean microscope slides, air dried, and placed on a slide warmer (37°C) for thorough drying. In our laboratory, we do not find it necessary to use adhesive coatings on the microscope slides, but we use specially treated histology slides (Fisher Scientific Superfrost/Plus™ *), as received from the supplier. Once dry, the slides can be stored indefinitely — if kept dry and protected from insects. To begin the staining process, the slides are placed vertically in a warming oven at 65 to 70°C for 1 hour to allow most of the wax to melt and drip off of the slides. After cooling to room temperature, the slides are brought to water saturation through a sequence of xylene (3 × 2 min); absolute ethanol (2 × 2 min); 95, 80, 50, and 50% ethanol, 2 min each; and finally distilled water. Some commonly used stains are described below.

2. Embedding and sectioning with plastics

A complete description of this procedure is outside the scope of this text, and the reader is referred to several texts on this topic. Examples are *Practical Methods in Electron Microscopy*, edited by Audrey M. Glauert, North Holland Publishing, Amsterdam (this is a series with 14 volumes through 1992 and covering most aspects of the preparation and sectioning of materials for both transmission and scanning electron microscopy); *Methods of Preparation for Electron Microscopy*, by D. G. Robinson et al., Springer-Verlag, New York, 1987.

* Registered trademark of Fisher Scientific Co.; Pittsburgh, PA.

3. Examples and applications of common tissue stains

Staining respiratory tissues for routine light microscopy usually employs readily available stains that are found in most histology or pathology laboratories. Those described below can be used for qualitative and quantitative analyses of major lung cell types by light microscopy. A very useful reference for other, less frequently used stains is Luna's *Manual of Staining Procedures of the Armed Forces Institute of Pathology*.[15] For a more in-depth discussion of stains, see references below.[16–20]

Hematoxylin and Eosin (H & E). This is probably the most frequently used tissue stain. It is easy to use, and it is adequate for routine qualitative analysis of lung pathology. H & E is rather nonspecific, in that cell nuclei are blue or purple, while other tissue components are shades of orange. Thus, cell differentiation is not always possible. Premixed H & E stains of good quality are available from scientific and hospital suppliers, so that it is not necessary to prepare the stains from scratch.

Materials

Hematoxylin, Gill's #2 (Fisher Scientific Co.; Pittsburgh, PA)
Eosin Y, 1% in 100% ethanol
Ammonia water: 4 drops 58% NH_4OH in 500 m*l* tap water

Procedure

If sections are not dewaxed, warm them in a laboratory oven (65 to 70°C) for 30 min, allow them to cool to room temperature, then soak them in two changes of xylene or toluene for 30 min. Bring dewaxed lung sections on slides to water saturation through an alcohol series: 100%, 80%, 50%, and two changes of distilled water. Drain well and immerse in hematoxylin for 3 min. Wash slides in gently running cold tap water 2 to 3 min until clear. Place in ammonia water for 1 min, then running tap water 1 min. Drain well, and place in eosin for 30 sec to 1 min. Check briefly with a microscope for intensity of staining. If stain is too intense, it can be removed by brief rinsing in a 70% ethanol in water solution. Dehydrate sections through alcohol to 100% xylene or toluene (10 dips each in 95% ethanol; three changes of 100% ethanol and 1:1 ethanol:xylene; then three changes of xylene, 3 min each). Mount using a #1 coverslip and Permount™,* DePex™,** or equivalent.

Alcian Blue-Periodic Acid-Schiff-Hematoxylin (AB-PAS-H). This stain is designed to stain mucopolysaccharides and thereby identify mucous cells. It can be used in quantitative studies to determine effects of inhaled pollutants on mucus production and proliferation of mucous cells in the airways.

* Registered Trademark of Fisher Scientific Co.; Pittsburgh, PA.
** Registered Trademark of BDH Laboratory Supplies; Poole, UK.

Materials

Alcian blue stock solution: 10 g Alcian blue 8GX (Color Index, C.I. 74240) per liter of water, plus 0.8 m*l* glacial acetic acid. pH should be approximately 2.5.

Alcian blue working solution: Dilute stock 1:10 in water acidified with 0.8 m*l* acetic acid per liter. Prepare just before use.

Periodic acid: 0.5% in water, made up just before use.

Reducing rinse: Recommended by some workers to remove traces of periodate which could cause nonspecific color development of the Schiff reagent. For each 100 m*l*:

> Potassium iodide (KI): 2 g
>
> Sodium thiosulfate ($Na_2O_3S_2$): 2 g
>
> Dissolve in 40 m*l* distilled water.
>
> Add, with continuous stirring, 60 m*l* absolute ethanol, then 1 m*l* of 2 *N* HCl. Allow any precipitate of sulfur to settle out.

Schiff reagent: Use commercial preparation (e.g., Schiff's Reagent; Sigma Chemical Co.; St. Louis, MO), or prepare as follows: Dissolve 4 g basic fuchsin (Pararosaniline chloride, C.I. 42500, Sigma Chemical Co.; St. Louis, MO) in 800 m*l* boiling water. Cool to room temperature, and add 80 m*l* 1 *N* HCl and 4 g sodium metabisulfite. Mix well, cover, and allow to stand overnight at room temperature. Add 2 g activated charcoal powder, mix very well for 1 to 2 min, filter through Whatman #1 paper, and store in a refrigerator. The filtrate should be water clear; if not, repeat the charcoal treatment. The solution is stable for at least several weeks if tightly closed and refrigerated. To test for proper staining activity, add a few drops to 10 m*l* of 37% formaldehyde. The solution should rapidly turn reddish purple. If the color development is slow and appears dark bluish-purple, the reagent should be discarded.

Sulfurous acid rinse: Prepare and use in fume hood because of evolution of SO_2 gas. For each 100 m*l*, mix 89 m*l* water, 6 m*l* of a 10% solution of sodium metabisulfite ($Na_2S_2O_5$), and 5 m*l* 1 *N* HCl.

Hematoxylin: Gill's #2 commercial preparation (Fisher Scientific Co.; Pittsburgh, PA).

Acid alcohol: 1 m*l* 1 *N* HCl per 100 m*l* 70% ethanol.

Ammonia water: 4 drops 58% NH_4OH in 300 m*l* water.

Procedure

Bring sections to water saturation as above. If the tissues were fixed in a fixative containing glutaraldehyde, the unreacted aldehyde groups must be reduced before proceeding. Prepare immediately before use a solution of 1% sodium borohydride ($NaBH_4$) in distilled water. Read and follow precautions on sodium borohydride container. Immerse slides for 5 min. A slight bubbling of hydrogen gas will occur due to decomposition of the $NaBH_4$. Rinse twice in distilled water and to complete the stain proceed as follows:

1. 30 min in Alcian blue working solution
2. 2 min cold running tap water
3. 5 min in periodic acid bath
4. 2 changes distilled water, 30 sec each
5. 5 min in reducing rinse (see discussion above)
6. 2 changes distilled water, 30 sec each
7. 15 min in Schiff reagent
8. 3 changes, 2 min each, in sulfurous acid rinse
9. 10 min cold running tap water
10. 1 min Gill's #2 hematoxylin
11. 1 dip in acid alcohol
12. 2 min cold running tap water or until clear
13. 10 dips in ammonia water
14. 10 dips in distilled water
15. Dehydrate through alcohols to xylene and mount as above

Result

The secretory granules in mucous cells will be bright red, nuclei will be bluish purple, and other structures will be shades of red and blue.

Collagen Stains. Staining for collagen and other connective tissues can be accomplished by several methods; however, the major drawback is their nonspecificity. One procedure which appears to overcome this difficulty involves the use of stains with a molecular structure that results in very stable binding to collagen molecules. One of these is Sirius red. This dye molecule associates with the collagen structure and gives a bright red stain when viewed using a light microscope, with little background color. The collagen-dye complex is also highly birefringent, which permits unequivocal localization of the collagen using a microscope with polarizing filters. The amount of dye bound per mole of collagen is not considered to be stoichometric, and it cannot be used to quantify collagen in tissue. However, if the staining procedure is carefully standardized and the tissue sections are of the same thickness, the relative amounts of collagen in tissues can be compared.

Materials

Sirius red F3BA (Direct Red 80, C.I. 35780, Roboz Surgical Instrument Co.; Rockville, MD)
Picric acid (Trinitrophenol; observe precautions, see Section II.A.1)
HCL, 0.01 *N*

Procedure[16]

1. Prepare a saturated solution of picric acid in water by adding 1.4 g (more, if necessary) to each 100 m*l* of water and stirring well. The solution is saturated when a few crystals remain after thorough mixing.
2. To each 100 m*l* of picric acid solution, add exactly 0.1 g of the Sirius red dye powder. Mix well and let stand overnight before use.
3. Dewax sections and bring to water as above.
4. Immerse for 1 hr in 0.01 *N* HCl.
5. Immerse for 1 hr in Sirius red solution.
6. Immerse for 5 min in 0.01 *N* HCl.
7. Drain slides briefly, and dehydrate rapidly by placing in 2 changes of 100% ethanol, followed by 2 changes of xylene.
8. Mount in Permount™ (Fisher Scientific Co.; Pittsburgh, PA) or equivalent.

Result

Collagen fibers stain bright red and show red birefringence with crossed polarizers. All other tissue elements are yellow.

Toluidine Blue. Plastic sections are usually cut much thinner than paraffin sections, and, hence, staining by H & E may not be intense enough for clear visualization of tissue structures. Toluidine blue staining can be controlled to give very intense staining of plastic sections that are less than 1 μm thick. Depending on the pH, varying degrees of metachromasia (a shift from blue to red) can be obtained to demonstrate mucopolysaccharides. The procedure given is for routine staining of plastic or paraffin sections.

Materials

Toluidine blue dye (Toluidine blue O, C.I. 52040, Sigma Chemical Co.; St. Louis, MO)
Sodium acetate: 1 *M* solution in water
HCl: 1 *N*

Procedure

1. For each 100 m*l*, dissolve 0.25 g of dye in a mixture of 20 m*l* 1 *M* sodium acetate, 14 m*l* 1 *N* HCl, and 66 m*l* water. Prepare at least 24 hr in advance of use. Adjust pH to 4.2 with 1 *N* NaOH or 1 *N* HCL, and filter before use.
2. Dewax slides, if necessary, and bring to water as above.
3. Immerse slides in stain up to 5 min. Follow progress of staining with microscope. Freshly prepared stain may require only 30 sec or less.
4. Rinse in 3 changes of distilled water.
5. Dehydrate through alcohols and xylene, and mount with Permount, DePex, or equivalent.

Result

Mucosubstances will be various shades of red to reddish-purple; other structures will be shades of blue.

References

1. Gehr, P., Geiser, M., Stone, K. C., and Crapo, J. D., Morphometric analysis of the gas exchange region of the lung, in Toxicology of the Lung, Second ed., Gardner, D. E., Crapo, J. D., and McClellan, R. O., Eds., Raven Press, New York, 1993, chap. 5.
2. Weibel, E.R., *Morphometry of the Human Lung,* Academic Press, New York, 1963, chap. IV.
3. Weibel, E.R., Morphometric and stereological methods in respiratory physiology including fixation techniques. Techniques in respiratory physiology — Part I, in *Physiology,* Vol. P4/1, Otis, A. B., Ed., Elsevier, New York, 1984.
4. Bracegirdle, B., *A History of Microtechnique: The Evolution of the Microtome and the Development of Tissue Preparation,* Heinemann, London, UK, 1978.
5. Borchard R. E., Barnes, C. D., and Eltherington, L. G., *Drug Dosage in Laboratory Animals: A Handbook,* Third ed., Telford Press, Caldwell, NJ, 1990.
6. Report of the AVMA Panel on Euthanasia, *J. Am. Vet. Med. Assoc.,* 202, 229, 1993.
7. Dungworth, D. L., Schwartz, L. W., Tyler, W. S., and Phalen, R.F., Morphological methods for evaluation of pulmonary toxicity in animals, *Ann. Rev. Pharmacol. Toxicol.,* 16, 381, 1976.
8. Hayatdavoudi, G., Crapo, J. D., Miller, F. J., and O'Neil, J. J., Factors determining degree of inflation in intratracheally fixed rat lungs, *J. Appl. Physiol.: Respir. Environ. Exercise Physiol.,* 48, 389, 1980.
9. Tylcr, W. S., Dungworth, D. L., Plopper, C. G., Hyde, D. M., and Tyler, N. K., Structural evaluation of the respiratory system, *Fund. Appl. Toxicol.,* 5, 405, 1985.
10. McClure, T. R., Diller, P. J., and Phalen, R. F., An airway perfusion apparatus for whole lung fixation, *Lab. Animal Sci.,* 32, 195, 1982.
11. Brody, A. R. and Craighead, J. E., A simple perfusion apparatus for lung fixation, *Proc. Soc. Exp. Biol. Med.,* 143, 388, 1973.
12. Login, G. R. and Dvorak, A. M., Methods of microwave fixation for microscopy. A review of research and clinical applications: 1970–1992, *Prog. Histochem. Cytochem.,* 27, 1, 1994.
13. Turner, C. R., Zuczek, S., Knudsen, D. J., and Wheeldon, E. B., Microwave fixation of the lung, *Stain Technol.,* 65, 95, 1990.
14. Scherle, W., A simple method for volumetry of organs in quantitative stereology, *Mikroskopie,* 25, 57, 1970.
15. Luna, L. G., Ed., *Manual of Histologic Staining Methods of the Armed Forces Institute of Pathology,* Third ed., McGraw-Hill, New York, 1968.
16. Chayen, J. and Bitensky, L., *Practical Histochemistry,* Second ed., John Wiley & Sons, Chichester, UK, 1991.

17. Gray, P., *Handbook of Basic Microtechnique,* Third ed., McGraw-Hill, New York, 1964.
18. van Noorden, C. J. F., *Enzyme Histochemistry: A Laboratory Manual of Current Methods,* Microscopy Handbooks #26, Oxford University Press, New York, 1992.
19. Clark, G., Ed., *Staining Procedures,* Fourth ed., Williams & Wilkins, Baltimore, MD, 1981.
20. Sanderson, J. B., *Biological Microtechnique,* Microscopy Handbooks #28, BIOS Scientific, Oxford, England, 1994.

Chapter 8

Specialized Sampling Techniques

Deepak K. Bhalla

Contents

0-8493-3343-1/97/$0.00+$.50

I. Introduction

Inhalation exposures of small laboratory animals are performed to understand the types and extents of biological changes or injuries produced in cells and tissues of the body by airborne pollutants. Since the cells lining the respiratory tract come in direct contact with inhaled pollutants, these cells constitute the primary targets. Therefore, studies in inhalation toxicology aimed at measuring biological responses often attempt to analyze the surfaces and cells of the respiratory tract. Analyses involving blood and its elements are performed to detect the extra-pulmonary effects. These analyses often require specific handling procedures and involve recovery of fluid and cells from the airway spaces and blood from the general circulation. The fluids generally require caution in handling and storage to preserve the biological activity of the samples. Similarly, the cells require appropriate manipulations, including centrifugations, enrichment, counting, and viability assessment, for their use in toxicological studies. This chapter therefore, is devoted to the specialized techniques that facilitate these selected biological measurements. The selection of these techniques was based on some of the well characterized effects of inhaled pollutants. For instance, air pollutants such as ozone, nitrogen dioxide, formaldehyde, and cigarette smoke produce leakiness of the airway walls, thereby increasing the transmucosal permeability,[1-4] they induce recruitment of inflammatory cells in the lung,[5-7] and they produce extrapulmonary effects,[8] presumably through the release of toxic products in the lung. Airway permeability measurements are made by either determining serum protein concentrations in the broncho-alveolar lavage fluid (BALF) or by introducing radiolabeled tracers into the lung, followed by their recovery from the blood. Inflammatory cells in airway spaces are studied following their isolation from the BALF, and the extrapulmonary effects can be detected by collecting blood and purifying the cells. The successful performance of these measurements is based on learning to perform a tracheostomy, collecting BALF, placing cannulas in the blood vessels, collecting blood samples, centrifuging fluids and suspensions, enriching cell preparations, counting cells and validating cell viability. The following sections describe these procedures.

II. Scope and Limitations

The injuries to various components of the lung following inhalation exposures to pollutants are detected by a variety of methods involving morphological, functional, and biochemical procedures. The morphological procedures involving tissue sampling are described in Chapter 7; this section primarily covers the procedures involving measurements of fluids. While there are numerous types of injuries detected by an even greater number of methodological approaches, the descriptions presented in this section are certainly not meant to include a broad spectrum of procedures; they are limited in their coverage to the commonly used procedures of interest to investigators studying lung fluid, blood, and isolated cells. It should also be emphasized at the outset that these are representative procedures, and they can be fine-tuned and customized for specific needs.

III. Safety Issues

Researchers working in laboratories are not only entrusted with maintaining high standards of conduct and organizational skills in performing specialized techniques, but they are also expected to maintain a safe environment for themselves and for the protection of their co-workers. Institutional policies also require work performance according to defined guidelines and the disposal of unused chemicals and other waste in compliance with institutional policies and applicable federal and state regulations. The goal of this section, therefore, is to summarize appropriate laboratory safety practices to be followed while performing research and to identify disposal procedures for hazardous chemicals and other substances encountered while performing sampling techniques discussed in this chapter.

Irrespective of the specific procedures performed, it is a common practice to wear apparel that covers most of the body, including the feet and legs, in order to avoid accidental spillage of a chemical on the skin. For the same reason, it is also a common practice for a worker to wear impervious gloves, a laboratory coat or apron, and sometimes a mask for work involving inhalants. A general recommendation for maintaining personal safety is to work in a clean, neat environment and to clean up spills quickly and efficiently. Precautions for handling hazardous substances generated in the course of performing techniques discussed in this chapter and appropriate disposal procedures are discussed below with specific examples.

Sharp items, such as razor blades, glass pipettes, syringe needles, etc., must be disposed of in special containers approved by the institution. These items should never be discarded in the general trash containers.

All radiochemicals should be handled with caution while wearing appropriate protective clothing. In addition, the counter surfaces should be covered with plastic-lined paper. In case of work involving isotopes that generate significant radiation, the workers may be required to obtain a radioisotope license, to wear radiation detecting badges, and to use lead foils, lead bricks, or thick plastic shielding. The radioactive waste is collected in specifically approved containers and is saved for pickup by authorized personnel.

Fixatives and hazardous nonradioactive chemicals and the generated waste should be stored in appropriate and approved containers. For example, flammables are kept in metal cabinets designed for their storage, and irritants and inhalants are stored and used in fume hoods. It is also a general rule that hazardous laboratory chemicals are not disposed of in water drains. They are usually stored in specific containers for disposal by trained officials of the institution. Because of the professional status and training of researchers, everyone working in a laboratory is expected to be responsible for maintaining safety for themselves, their colleagues, and the general public.

IV. Surgical Procedures

Surgical procedures necessary to acquire needed samples are described below.

A. Anesthesia

It is important that the animals be handled carefully and cared for according to the guidelines set by federal agencies, state laws, and institutional committees. It is also an accepted laboratory practice to perform the essential surgical procedures using standard guidelines so as to minimize pain and discomfort to the animals during the invasive procedures. Therefore, it is important that the animals be adequately anesthetized and not simply paralyzed. A general anesthetic in common use for small animal surgery is sodium pentobarbital

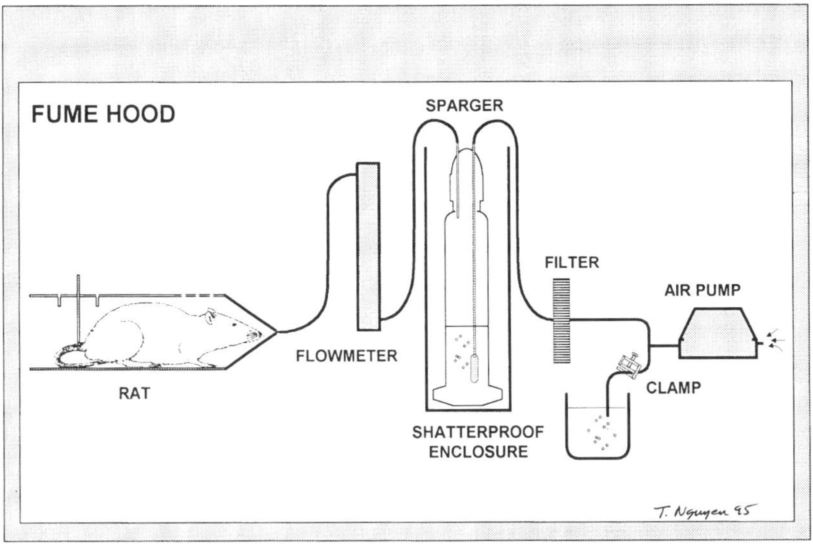

Figure 8.1
Setup for methoxyfluorane inhalation by small animals. Safety requirements include use of a ventilated hood, enclosures for glass sparger, and a pressure-relief system (bubbler in the diagram).

(nembutal sodium). An adequate dose for rats is about 5 to 10 mg per 100 g body weight. This dose is achieved by diluting the stock solution (64.8 mg/m*l*) of sodium pentobarbital 1:2 with sterile saline. This mixture is mixed well in a clean glass vial, and then 1 m*l* is injected intraperitoneally (i.p.) into a 250- to 300-g rat using a 26-gauge needle. Rats may be restrained during the injection by using commercially available plastic restraining tubes. The rats are generally sedated and ready for surgery in 3 to 5 minutes, but dosage will vary for individual strains of rats. This protocol has been found adequate for maintaining rats under anesthesia for at least 1 hr.[9] For procedures requiring rapid anesthesia and a rapid recovery, an anesthetic of choice is the inhalant methoxyfluorane (Metofane).[10] The anesthetic is generated in a fume hood by placing the chemical in a sparger and bubbling filtered air through it (Figure 8.1). A safety outlet fitted with a clamp is used to release air and regulate air pressure. Vapor generated in this manner is supplied through a nose-cone to a rat. During anesthesia delivery, the rat is restrained in a plastic tube fitted with a tail-gate that prevents the rat from backing out of the tube. The flow of vapor is regulated using a flowmeter placed between the sparger and the nose-cone.

B. Surgery

1. Tracheostomy

The anesthetized rats are placed in a supine position on a surgical board and held in place by taping their forearms and hind legs to the board. The thoracic

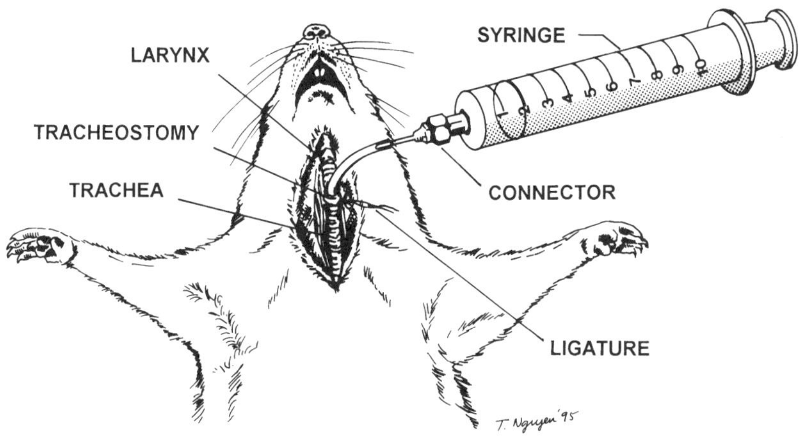

Figure 8.2
Tracheostomy preparation for instillation or lavage.

region is prepared by wetting it with 70% alcohol from a wash bottle. A skin incision, made in the thoracic region, is extended from the neck to the base of the anterior end of the abdomen using medium-sized scissors. Another incision is made in the muscular tissue over the trachea using small scissors or a scalpel. Following this incision, the trachea is exposed by inserting curved forceps under the trachea and lifting it slightly over the adjoining tissues. For tracheal cannulation, an incision is made in the trachea between the third and fourth tracheal rings. A polyethylene tube, 0.050" i.d. × 0.09" o.d., is beveled and inserted through the tracheal incision and advanced to reach approximately the sixth tracheal ring. The cannula is then tied in place by inserting a suture underneath the trachea and tying the two ends over the ventral surface of the trachea (Figure 8.2).

2. Vascular catheterization

The process of placing a canula into a femoral artery or vein is similar to that involving placement of a tracheal canula, but it is a more delicate procedure that requires greater care. Perfection of this technique will require patience, skill, and practice. In preparation for the surgery, cannulas are prepared by cutting approximately 9" segments of PE-10 tubing. The tubing is carefully slipped over a 27-gauge needle and the free end is beveled. The needle is then attached to a 1-m*l* syringe, and 0.2 m*l* of a sterile solution containing 20 units of heparin in saline is drawn into the syringe. The femoral veins and arteries run side by side into the hind legs. The exposure of these blood vessels requires incision of the skin, removal of the muscular tissue, and separation of the membranous sheaths enclosing the two blood vessels. All this can be done using small scissors and two pairs of curved forceps. After separation of the artery and the vein, each blood vessel is lifted up using curved forceps. While holding this position, a fine incision is made by inserting a 23- to 26-gauge

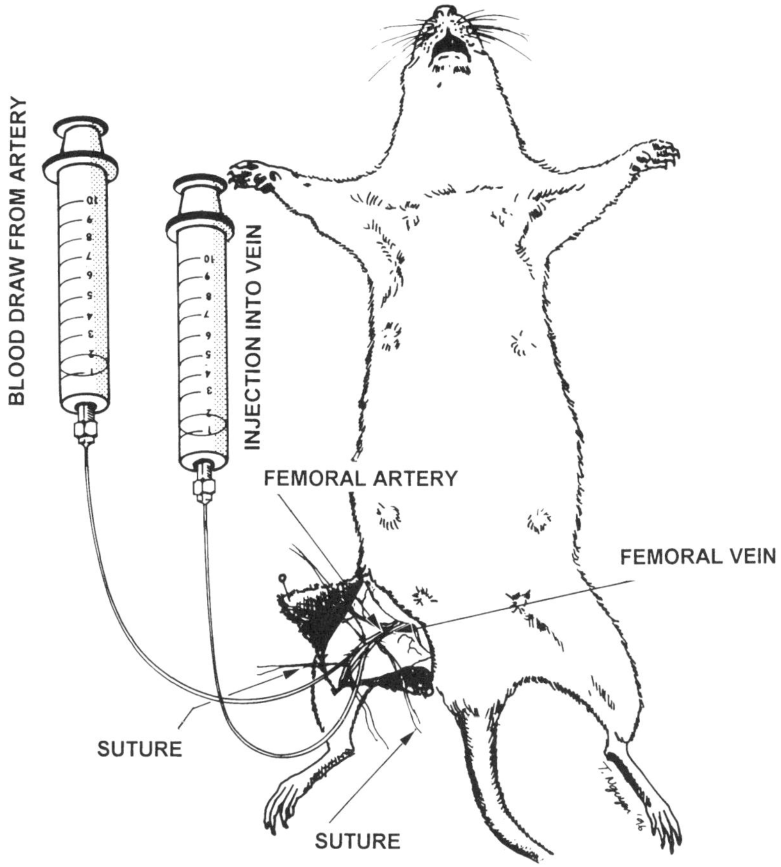

Figure 8.3
Femoral cannula preparation for anesthetized rat.

needle into the blood vessel. The PE-10 tubing attached to the heparin-filled syringe is then inserted into the blood vessel through the incision and tied in place using a suture (Figure 8.3). Another suture is used to tie the open end of the blood vessel distal to the cannulation site. This prevents drainage of blood from distal body parts supplied by the blood vessels.

C. Practical Applications of Tracheostomies and Vascular Cannulas

Tracheostomies are performed to introduce test agents into the lungs[11] or for obtaining fluid[12,13] or cellular components[14-18] from the lung. For intratracheal instillations, care is exercised to minimize interference with normal breathing of the animal while introducing the inoculum. Therefore, the test agents are dissolved in a small volume of a physiological buffer and introduced slowly through the tracheostomy tube. For practical purposes, 0.1 m*l* of inoculum can be safely introduced over a period of 5 min or less.

Vascular cannulas can be used for intravascular injections, continuous infusion of test agents into a vein, or timed collection of blood samples from an artery.[18] Initially, a small amount of blood is drawn into the syringe to confirm proper placement of the cannula, and then heparin is introduced into the blood vessel to clear the cannula and prevent further clogging of the cannula due to blood clotting. Test samples are then introduced using a 27-gauge needle inserted into the PE-10 tubing, either manually or using an infusion pump for continuous delivery over a period of time.

D. Equipment and Supplies

Item	Typical supplier
Na-pentobarbital injection, 64.8mg/m*l*	Western Medical Supply
Methoxyfluorane (Metaphane)	Pitman-Moore, Inc.
23-guage × 1" needles	Monoject
3-cc syringes, luer-lok	Becton Dickinson
Physiological saline injection, 0.9%	McGaw
Pan balance, XE series, Model 3000D	Denver Instrument Co.
Rat restrainer	Fisher
Tissue forceps, 1 × 2 teeth, 4-1/2", #RS-8160	ROBOZ
Micro-dissecting forceps, smooth jaws, 6" angular, #RS-5359	ROBOZ
Rankin-Kelly forceps, 6-1/2", straight, #RS-7140	ROBOZ
Operating scissors, 5", straight, BB, #RS-6810	ROBOZ
Micro-dissecting scissors, 4 1/2", straight, blunt, #RS-5942	ROBOZ
Suture, 00 silk, code t-55	Ethicon
8/3 polyester, white bonded, #421	
2" × 2" cloth gauze squares	Johnson & Johnson Medical
Ethanol, 75%	Fisher

V. Bronchoalveolar Lavage

Lungs of anesthetized rats are usually lavaged with an isotonic buffer to recover either free cells or lung lining fluid.

A. Lavage Buffers

The buffers used for lavage are usually modified, depending upon the specific research need. For example, buffers containing Ca^{++} and Mg^{++} are not ideally suited for the recovery of cellular contents; therefore, these divalent anions are omitted from the buffer when cell recovery is the focus.

Commercially available Hank's balanced salt solution (HBSS), with or without Ca^{++} and Mg^{++}, is often used for performing bronchoalveolar lavage. Phosphate buffered saline (PBS), another buffer found suitable for lavage, can be purchased or easily prepared in the laboratory by dissolving individual components in distilled water. The various constituents and their ratios for preparing 1 liter of the buffers are as follows.

1. Hank's balanced salt solution with Ca^{++}, Mg^{++}, and HEPES

For preparation of 1 liter, one will need:

100 m*l* 10× HBSS
0.1 g $MgSO_4 \cdot 7H_2O$
0.35 g sodium bicarbonate
5.96 g HEPES

Mix above components in 800 m*l* distilled water. Separately, mix 0.185 g $CaCl_2$ in 100 m*l* water, then add *slowly* to the above. Adjust pH to 7.2 with 1 *N* NaOH and add water to 1 liter volume (pre-made solution can be ordered through Life-Technologies).

2. Hank's balanced salt solution without Ca^{++} or Mg^{++}

For preparation of 10 liters, one will need:

4 g KCl
0.6 g KH_2PO_4
80 g NaCl
0.9 g $Na_2HPO_4 \cdot 7H_2O$
10 g glucose
0.1 g Phenol red (optional)

Mix the above compounds in 900 m*l* distilled water, then add water to 1 liter volume. To use, add 0.35 g sodium bicarbonate per liter (pre-made solution can be ordered through Life-Technologies)

3. Phosphate buffered saline

1000 m*l* distilled water
1.15 g sodium phosphate ($Na_2HPO_4 \cdot 7H_2O$)
0.20 g potassium phosphate (KH_2PO_4)
8.0 g sodium chloride (NaCl)
0.20 g potassium chloride (KCl)
0.10 g calcium chloride ($CaCl_2$)
0.10 g magnesium chloride ($MgCl_2 \cdot 6H_2O$)

All of the contents of the buffer, except $CaCl_2$, are mixed in 900 m*l* of distilled water. Calcium chloride is mixed separately in the remaining 100 m*l* of distilled water and added slowly to the rest of the buffer after all other components are dissolved. This prevents cloudiness of the buffer, which can happen if $CaCl_2$ powder is added directly to the solution.

B. Lavage Procedure

Immediately prior to starting the lavage, the diaphragm of the rat is cut to permit free expansion and collapse of the lung lobes during lavage. A major blood vessel, usually the dorsal aorta, is severed to eliminate the continuous blood supply to the lung. Lavage is then performed by introducing an appropriate buffer into the lungs through the tracheostomy tube while the animal is under anesthesia. Generally, 7 to 10 m*l* of the buffer is infused into the trachea and then withdrawn while holding the syringe in place. This process of introduction and withdrawal is repeated three times and the fluid is then collected in a vial. The collected samples can be stored for a short period at 4°C, but a low-temperature freezer (i.e., –70°C) is recommended for long-term storage to minimize enzyme degradation.

The lavage procedure for recovery of monocytes and neutrophils is essentially the same as described above, with the exception that a Ca^{++}-free medium is used and the process is repeated with two additional aliquots of the buffer to maximize cell recovery. After final withdrawal, all of the samples are combined and the suspensions are centrifuged at 200× *g* for 10 min to recover the cells in a pellet. The supernatant is discarded and the cell pellet is resuspended in a desired volume of the buffer.

C. Equipment and Supplies (with a Typical Supplier)

In addition to the surgical equipment listed above:

Item	Typical supplier
Tracheostomy tubes, polyethylene, 0.05" i.d. × 0.09"-gauge o.d.	Cole-Parmer
18-gauge × 1-1/2" needle	Monoject
Centrifuge tubes	Costar
Ice bucket	Fisher
HBSS without Ca^{++} or Mg^{++}, with HEPES at 37°C	Life Technologies
Water bath	Fisher
Nalgene® cryovials, 2.0 m*l*, #03-337-7D	Fisher
Nalgene® cryoboxes, 9 × 9 array, clear lid, #15-350-107B	Fisher

VI. Neutrophil Isolation from Whole Blood

Neutrophils are generally isolated from the blood. For this purpose, the rats are deeply anesthetized with 250 mg/kg body weight of sodium pentobarbiatal (an overdose), the abdomen and thoracic cavities are surgically opened, and the blood is drawn either directly from the heart or the aorta using an 18-gauge needle attached to a 10-m*l* syringe. Two procedures have been used in our laboratory to further process the blood for neutrophil isolation. Both of these procedures are described in order to accommodate individual needs. Selection of a particular procedure will depend on individual preference, convenience of the procedure, and retention of cellular activity or biological functions of interest to the investigator.

A. Isolation Procedure

Neutrophil isolation from the blood can be accomplished by a combined sedimentation and density centrifugation procedure.[20] About 8 m*l* of blood are collected from each animal via cardiac puncture into a syringe containing 2 m*l* of heparinized HBSS without Ca^{++} or Mg^{++}. The total volume is then placed in a silicone-coated (Sigmacote, Sigma Chemical Co.; St. Louis, MO) glass test tube and an equal volume of dextran solution (6% dextran-500 kd in HBSS with HEPES and heparin, Sigma Chemical Co.; St. Louis, MO) is added and mixed by inversion. Red blood cells are allowed to sediment for up to 1 hour, after which the leukocytes in the supernatant are removed and placed in a 50-m*l* silicone-coated glass test tube. The leukocytes are slowly underlayered with approximately 10 m*l* of Histopaque-1077 (Sigma Chemical Co.; St. Louis, MO), and centrifuged at 400× *g* for 30 min. The lymphocyte layer formed at the Histopaque-HBSS interface is carefully removed, and neutrophils in the pellet at the tube bottom are resuspended. The cell fraction is washed once with HBSS, centrifuged, and resuspended in 1 m*l* HBSS. Remaining red blood cells are lysed by adding 9 m*l* of a lysing solution (consisting of 0.15 *M* NH_4Cl + 0.01 m*M* $NaHCO_3$ + 0.001 *M* Na_2EDTA). After centrifuging at 200× *g* for 5 min, the cell pellet is resuspended in the media and counted using either a hemacytometer, as described in Section VII.A, or in a Coulter cell counter (CBC 5, Coulter Electronics, Inc.; Marietta, GA), if available. It is a good practice to determine the cell viability by Trypan blue exclusion using a bright-line hemacytometer (see Section VII.B for the procedure). A cytometry slide can be made for the neutrophil preparations and stained with Diff Quik (Baxter Healthcare Corp.; Glendale, CA) for cell type analysis. The entire isolation procedure is completed in 8 to 10 hr.

A second procedure for neutrophil isolation, which takes 1 to 2 hr for completion, relies on a specific density gradient medium (NIM™-2, supplied by Cardinal Associates; Santa Fe, NM). The procedure recommended by the supplier takes advantage of an original observation that leukocytes from whole blood can be separated using a combination of sucrose polymer and sodium diatrizoate.[20] The isolation involves a two-component step gradient which is used to separate mononuclear cells and polymorphonuclear leukocytes in a single centrifugation. In preparation for neutrophil isolation, bring the NIM™-2 to room temperature and mix it by inverting the bottle several times. Add 2.5 m*l* of NIM™-2A into a test tube, and then slowly layer 2.5 m*l* of NIM™-2B on top of NIM™-2A, taking care that the two solutions do not intermix. To separate neutrophils, collect EDTA-treated blood in a test tube. Since heparin was found to produce poor results, EDTA is recommended by the supplier as a preferred anticoagulant. If storage is necessary, the blood should be kept at 2 to 8°C; however, for optimal results, separations within 2 hours of collection are recommended. If the blood is stored at low temperature, bring it to room temperature and carefully layer 3 m*l* of blood on top of NIM™-2B. Place equally balanced tubes with NIM™-2 and blood in a swinging bucket rotor and centrifuge at 900× *g* for 45 min at 20°C. Let the rotor coast to full stop without applying the brakes. While the plasma forms the top layer, mononuclear cells are aggregated as a cloudy layer at the interface of plasma and the upper gradient; the neutrophils form a band at the interface of upper and lower gradients; and the red blood cells form a pellet at the bottom (Figure 8.4). Discard the plasma and collect the neutrophils into another tube containing Ca^{++}- and Mg^{++}-free HBSS. Mix the cells with HBSS by inverting several times and centrifuge at 400× *g* for 10 min. Discard the supernatant, resuspend the cells into 2 m*l* of the erythrocyte lysing solution (E-lyse), and centrifuge at 400× *g* for 5 min. Again, discard the supernatant, wash once again with 5 m*l* of HBSS, resuspend the isolated neutrophils in a desired volume of the buffer, count using a hemocytometer, and evaluate cell viability as described in Section VII.B.

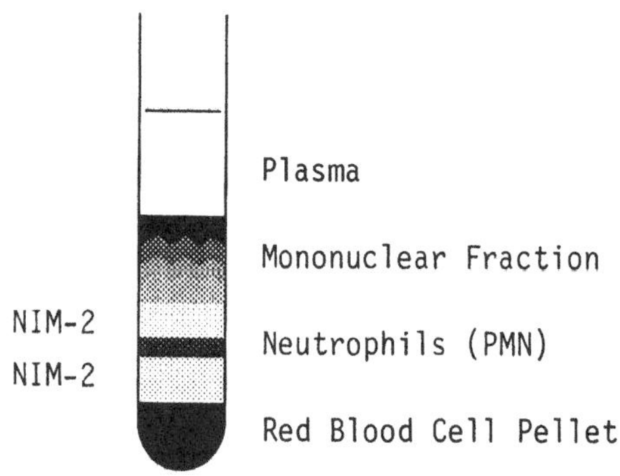

Figure 8.4
Gradient tube. (Courtesy of Cardinal Associates; Santa Fe, NM.)

B. Equipment and Supplies

Item	Typical supplier
NIM™-2 system	Cardinal Assoc., Inc.
NIM™-2, #236-105	
HBSS, #255-108	
E-Lyse, #255-103	
Centrifuge, refrigerated	I.E.C.
HyLite cell counting chamber, #02-671-10	Fisher
Heparin sodium injection, 1000 units/m*l*	Western Medical Supply
Test tubes, 13 × 100 mm, borosilicate	Fisher or VWR
Serological pipettes and bulb	Fisher or VWR
Pasteur pipettes and bulb	Fisher or VWR
3-cc syringes	Fisher or VWR
23-gauge needles	Fisher or VWR
Parafilm	Fisher or VWR

VII. Cell Counting, Viability Assessment, and Slide Preparation

When working with isolated cells, it is often necessary to adjust the cell number in a certain volume of the working solution. It is also essential that the cells isolated under experimental conditions remain viable to permit measurement of their biological functions. Therefore, procedures for counting cells and measuring their viability constitute an essential component of studies utilizing isolated cells.

A. Cell Counting

A common method of cell counting utilizes a hemocytometer. This apparatus consists of two chambers which are covered by a thin glass coverslip. Under the microscope, each chamber is seen as marked by five squares, four in the corners and one in the middle (marked 1 to 5 in Figure 8.5). Each one of the five squares is further divided by the grids and represents a region equivalent in volume to 1×10^{-4} cm^3. Cells suspended in a buffer, such as phosphate buffered saline or HBSS, are drawn into a Pasteur pipette and transferred to one of the chambers of the hemocytometer in such a way that the chamber is completely filled. When the Pasteur pipette containing the cells is placed at the edge of the coverslip covering the hemocytometer, and a gentle pressure is applied to the pipette bulb, the liquid containing the cells is drawn into the chamber under the coverslip by capillary action. Overfilling or underfilling of the chamber can introduce an error. The cells are counted (using a 10× objective and a 10× eyepiece) in each square of the hemocytometer chamber. This

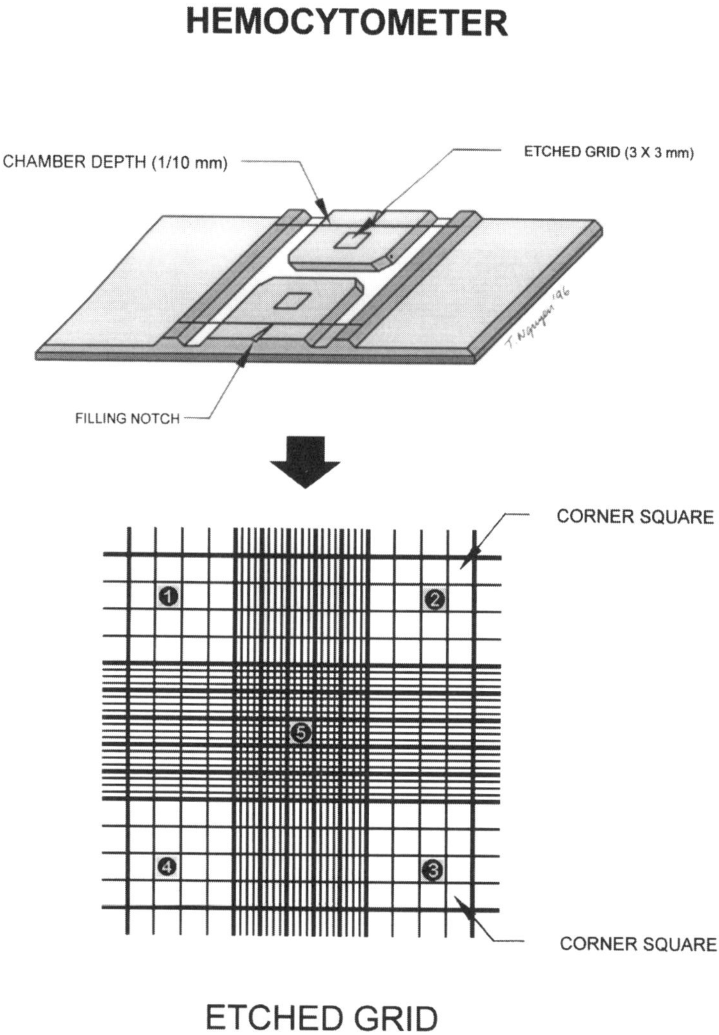

Figure 8.5
Hemocytometer with detail of etched grid.

count represents the number of cells in 1×10^{-4} cm^3. Since the cells are unstained at the time of counting, use of a phase-contrast microscope greatly facilitates cell identification. The total cell count is then determined by the following formula:

$$\frac{\text{Number of cells} \times \text{dilution factor} \times 10^4}{\text{Number of squares counted}} \quad (8.1)$$

B. Cell Viability

Trypan blue exclusion is widely accepted as an indicator of cell viability. The test is based on the fact that intact viable cells are impermeable to Trypan blue, whereas damaged or dead cells are permeable to the dye. The dead or nonviable cells are then recognized as stained cells. To perform the test, transfer the cell suspension in a volume of 0.5 m*l* of buffered solution (e.g., HBSS) to a small test tube. Add an equal volume of 0.4% Trypan blue to the cell suspension and mix well. After about 5 min, transfer a small volume of the cell suspension to one of the chambers of a hemocytometer so as to completely fill the chamber, and then count all the stained and unstained cells in the four corner squares. Determine the percent cell viability by the following formula:

$$\frac{\text{Number of unstained cells} \times 100}{\text{Total cell number (stained plus unstained)}} \tag{8.2}$$

C. Equipment and Supplies

Item	Typical supplier
HyLite counting chamber, #02-671-10	Fisher
Trypan blue, 0.4%, #630-5250AE	Life Technologies
Test tubes, 75 × 100 mm borosilicate	Fisher
Micropipettor, Gilson "Pipetteman", or	Ranin
Oxford "Benchmate"	Fisher
Pipette tips, Redi-tip	Fisher or Continental Lab Prod.
Laboratory counter, 5-key	Fisher
Microscope	Nikon

D. Slide Preparation

If the purpose of collecting cells, as in the lavage procedure, is to identify cell types for determining cell number changes, then the cells should be collected on a slide, fixed, stained, and viewed under a microscope. This is accomplished using either a "cytospin" device or a centrifuge equipped with a special device to collect cells onto a slide. Such a device, consisting of a "cytobucket" and associated accessories, is sold by International Equipment Company (Needham, Heights, MA). Prior to collecting cells, clean slides with a detergent, wash them thoroughly in running water, follow with a wash in distilled water, and allow to dry. Store in a dust-free box. Next, assemble the slides in the cytobuckets by first inserting a foam pad, provided by the supplier, into the holder. Code the slide and place it on the foam pad. After installing the slide, place the cell block on the slide and tighten in place using a clamp and tightening screws gently. Make sure that slides are secure, but not overly tightened. Now add 0.2 m*l* of cell suspensions onto the slides through the holes on top of the cell block. In this manner, cells can be added into each of the three chambers formed by the placement of the cell block on the slide. Centrifuge the cells in the cytobuckets using a horizontal rotor at 200× *g* for 10 min. After cells are deposited on the slides, use a Pasteur pipette to remove supernatant from the outlet hole and fix cells by adding 95% alcohol through the outlet hole and removing it after a few seconds. Remove the assembly by unscrewing the clamp, lift the cell block, remove the slide, and let the cells air-dry. The cells deposited on the slide are routinely stained in our laboratory using Diff-Quick and are viewed in a light microscope for differential cell analysis.

E. Equipment and Supplies

Item	Typical supplier
Cytobuckets, Trunnion carriers, #1024	I.E.C.
Glass microscope slides, SuperFrost Plus	Fisher
Coplin jar, glass with ground glass lid	Fisher
Pipettor and tips (see above)	Fisher
Centrifuge, CentralMP4R	I.E.C.
Methanol	Fisher
Diff-Quik stain set, #B4132-1	Baxter
90° crosshair eyepiece micrometer insert, #12-578-9	Fisher

VIII. Problems and Solutions

A. Procedure: Anesthesia

Problem: Breathing difficulty following anesthesia but prior to surgery.
Solution: Animals left on wood chips following anesthesia sometimes bury themselves in the bedding. As a result, they may inhale small pieces of the chips that get trapped in the nose and cause breathing difficulties. This can be prevented by placing the animals in a clean cage following anesthesia. In case of inadvertent blockage of nasal passages, the animal may be provided air through the tracheostomy tube.

B. Procedure: Surgery

Problem: Breathing difficulties during tracheal instillations.
Solution: Decrease the total volume and rate of the inoculum introduced at a given time. It is important that inoculum be instilled slowly so that it does not interfere with normal breathing. The bubbling of inoculum, as noticed by

regular rise and fall of the fluid meniscus in the tracheostomy tube, is due to forces exerted on the tracheal fluid by breathing and is normal.

C. Procedure: Bronchoalveolar Lavage

Problem: Difficulty in recovering lavage fluid.
Solution: This could happen if the needle used on the syringe for collecting lavage fluid is too small (e.g., gauge 26). Generally, an 18-gauge needle is used for performing bronchoalveolar lavages.

D. Procedure: Cell Viability

Problem: Cells recovered by bronchoalveolar lavage are not viable.
Solution: This could happen if the buffer used for lavage is not isotonic or if the cells are left at room temperature for too long. Checking the buffer for recipe accuracy and keeping the cells in an ice bucket for short-term storage will alleviate this problem.

E. Procedure: Bronchoalveolar Lavage

Problem: Blood in the lavage fluid.
Solution: Surgical exposure of the trachea is sometimes accompanied by rupture of small blood vessels. Blood from damaged vessels can be removed by using gauze pads prior to performing the tracheostomy to prevent entry of blood into the trachea during placement of the tracheostomy tube. Care should also be taken to avoid damaging lung lobes while making an incision in the diaphragm. Appearance of blood in the lavage fluid due to lung damage could seriously compromise the experiment.

F. Procedure: Surgery

Problem: The animal suddenly dies following heparin introduction into the femoral artery.
Solution: Make sure that the heparin in the syringe is free of bubbles, as the trapped bubbles could cause a fatal air embolism.

G. Procedure: Blood Sampling

Problem: Blood clots form in the vascular catheter, thereby making it difficult to draw blood.
Solution: Check to determine if the heparin has passed the expiration date. Generally, a repeat injection of heparin will dislodge the clots and keep the

tubing clear. If the blood draws are to be performed at intervals over a period of more than a few minutes, it is recommended that the catheter be cleared by introducing fresh heparin following each blood draw.

H. Procedure: Cell Counting

Problem: Difficulty with macrophage counting due to the presence of a large number of red blood cells in the lavage fluid.
Solution: Lyse the red blood cells by hypotonic shock (immersing the cells in distilled water for a few seconds), or use a lysing solution as described in the procedure for neutrophil isolation in Section VI.A.

I. Procedure: Cell Viability

Problem: The background is too dark after staining with Trypan blue.
Solution: Trypan blue binds to serum proteins. Washing the cells with a serum-free buffer, followed by centrifugation and resuspension in fresh medium prior to addition of Trypan blue, will eliminate background staining.

J. Procedure: Cell Viability

Problem: Most or all of the cells are stained, but you suspect the cells to be viable.
Solution: When cells are left in contact with trypan blue for a long period, even viable cells pick up the dye, thereby giving inaccurate results. Do not leave the cells in contact with the dye for an extended period.

K. Procedure: Cytospin Slides

Problem: Cell suspension leaks through the cell block when placed on slides.
Solution: Form a better seal between the cell block and the slide by applying a thin coat of silicone vacuum grease to the underside of the cell block before placing it on the slide.

L. Procedure: Cell Preparation

Problem: Cells do not adhere to the slide.
Solution: Either add 0.2% bovine serum albumin to the cell suspension or apply a thin coat of Mayer's albumin to the slide before placing it in the cytobucket.

IX. Suppliers

The list of equipment suppliers given below was prepared on the basis of author's familiarity and the use of equipment in his laboratory. The list is neither comprehensive nor does it endorse a particular manufacturer.

A.G. Heinze, Inc.
3 Watson
Irvine, CA 92718-2767

Baxter Scientific Products
17111 Redhill Avenue
Irvine, CA 92714

Becton Dickinson & Co.
2350 Qume Drive
San Jose, CA 95131

Brinkmann Instruments Co.
(A Division of Sybron Corp.)
1 Cantiague Road
Westbury, NY 11590

Cardinal Associates, Inc.
P.O. Box 5520
Santa Fe, NM 87502

Cole-Parmer Instruments Co.
625 East Bunker Court
Vernon Hills, IL 60061

Continental Lab Products
9240 Mira Este Court
San Diego, CA 92126

Costar
1 Alewife Center
Cambridge, MA 02140

Denver Instrument Co.
6542 Fig Street
Arvada, CO 80004

Ethicon, Inc.
Johnson & Johnson
P.O. Box 6800
Pistcaway, NJ 08855-6800

Fisher Scientific
711 Forbes Avenue
Pittsburgh, PA 15219-4785

International Equipment Co.
(A Division of Damon)
300 2nd Avenue
Needham Heights, MA 02194

Johnson & Johnson Health Care Systems
P.O. Box 6800
New Brunswick, NJ 08903

Life Technologies Inc.
3175 Staley Road
Grand Island, NY 14072

Shandon Lipshaw, Inc.
171 Industry Drive
Pittsburgh, PA 15275

McGaw, Inc.
2525 McGaw Avenue
Irvine, CA 92714

Organon Teknika
Cappel Division
100 Akzo Avenue
Durham, NC 27712

Pierce Chemical Co.
3747 N. Meridian Avenue
Rockford, IL 61101

Pitman-Moore
421 East Hawlay
Mundelein, IL 60060

Plas Labs
917 E. Chilson Street
Lansing, MI 48906

Ranin Instruments Co.
5400 Hollis Street
Emeryville, CA 94608

ROBOZ Surgical Instrument Co., Inc.
9210 Corporate Blvd., Suite 220
Rockville, MD 20850

Sigma Chemical Co.
P.O. Box 14508
St. Louis, MO 63178

Tecan, U.S., Inc.
P.O. Box 13953
4022 Stirrup Creek Road
Durham, NC 27703

Western Medical Supply, Inc.
117 E. Huntington Drive
Arcadia, CA 91006

References

1. Crocker, T. T. and Bhalla, D. K., Transport of macromolecules and particles at target sites for deposition of air pollutants, Health Effects Institute, Cambridge, MA, *HEI Rep.*, 3, 15, 1986.
2. Ranga, V., Kleinerman, J., Ip, M. P. C., and Collins, A. M., The effect of nitrogen dioxide on tracheal uptake and transport of horseradish peroxide in the guinea pig, *Am. Rev. Respir. Dis.*, 122, 483, 1980.
3. Jones, J. G., Lawler, P., Crawley, J. C. W., Minty, B. D., Hulands, G., and Veall, N., Increased alveolar epithelia permeability in cigarette smokers, *Lancet*, 1, 66, Jan., 1980.
4. Hogg, J. C., Pare, P. D., and Boucher, R. C., Bronchial mucosal permeability, *Fed. Proc.*, 38, 197, 1979.
5. Young, C. and Bhalla, D. K., Effects of ozone on the epithelial and inflammatory responses in the airways: role of tumor necrosis factor, *J. Tox. Environ. Health*, 46, 329, 1995.
6. Holtzman, J. J., Fabbri, L. M., O'Byrne, P. M., Gold, B. D., Aizawa, H., Walters, E. H., Alpert, S. E., and Nadel, J. A., Importance of airway inflammation for hyperresponsiveness induced by ozone, *Am. Rev. Respir. Dis.*, 127, 686, 1983.
7. Koren, H. S., Devlin, R. B., Graham, D. E., Mann, R., McGee, M. P., Horstman, D. H., Kozumbo, W. J., Becker, S., House, D. E., McDonnell, W. F., and Bromberg, P. A., Ozone-induced inflammation in the lower airways of human subjects, *Am. Rev. Respir. Dis.*, 139, 407, 1989.
8. Bhalla, D. K., Rasmussen, R. E., and Daniels, D., Adhesion and motility of polymorphonuclear leukocytes isolated from the blood of rats exposed to ozone: potential biomarkers of toxicity, *Tox. Appl. Pharm.*, 123, 177, 1993.

9. Bhalla, D. K. and Crocker, T. T., Tracheal permeability in rats exposed to ozone: an electron microscopic and autoradiographic analysis of the transport pathway, *Am. Rev. Respir. Dis.,* 134, 572, 1986.
10. Baker, C. E., Methoxyfluorane, in *Physician's Desk Reference,* 36th ed., Medical Economics Data, Montvale, NJ, 1982, 558–559.
11. Bhalla, D. K., Rasmussen, R. E., and Tjen, S., Interactive effects of O_3, cytochalasin D and vinblastine on transepithelial transport and cytoskeleton in rat airways, *Am. J. Respir. Cell Mol. Biol.,* 3, 119, 1990.
12. Schlesinger, R. B., Weideman, P. A., and Zelikoff, J. T., Effects of repeated exposures to ozone and nitrogen dioxide on respiratory tract prostanoids, *Inhal. Toxicol.,* 3, 27, 1991.
13. Gunnison, A. F., Finkelstein, I., Weideman, P., Su, W., Sobo, M., and Schlesinger, R. B., Age-dependent effect of ozone on pulmonary eicosanoid metabolism in rabbits and rats, *Fund. Appl. Toxicol.,* 15, 779, 1990.
14. Canning, B. J., Hmieleski, R. R., Spannhake, E. W., and Jakab, G. J., Ozone reduces murine alveolar and peritoneal macrophage phagocytosis: the role of prostanoids, *Am. J. Physiol.,* 261, L277, 1991.
15. Kleinman, M. T., Bhalla, D. K., Ziegler, B., Bucher-Evans, S., and McClure T., Effects of inhaled fine particles and ozone on pulmonary macrophages and epithelia, *Inhal. Toxicol.,* 5, 371, 1993.
16. Hotchkiss, J. A., Harkema, J. R., Kirkpatrick, D. T., and Henderson, R. F., Response of rat alveolar macrophages to ozone: quantitative assessment of population size, morphology, and proliferation following acute exposure, *Exp. Lung Res.,* 15, 1, 1989.
17. Bassett, D. J. P., Bowen-Kelly, E., Brewster, E. L., Elbon, C. L., Reichenbaugh, S.S., Bunton, T., and Kerr, J.S., A reversible model of acute lung injury based on ozone exposure, *Lung,* 166, 355, 1988.
18. Driscoll, K. E. and Schlesinger, R. B., Alveolar macrophage-stimulated neutrophil and monocyte migration: effects of in vitro ozone exposure, *Toxicol. Appl. Pharmacol.,* 93, 312, 1988.
19. Bhalla, D. K., Mannix, R. C., Kleinman, M. T., and Crocker, T. T., Relative permeability of nasal, tracheal and bronchoalveolar mucosa to macromolecules in rats exposed to ozone, *J. Toxicol. Environ. Health,* 17, 269, 1986.
20. Boym, A., Isolation of leucocytes from human blood, *Scand. J. Clin. Lab. Invest. Suppl.,* 21(97), 77, 1968.

Index

A

B

E

F

J

K

L

M

N

O

P

Q

R

S

T

U

V

W

Z